# Vue.js
## 光速入门及企业项目开发实战

庄庆乐 任小龙 陈世云 编著

清华大学出版社
北京

## 内 容 简 介

本书采用简洁直观的方式来讲解 Vue.js 2.0 的各方面，并融入了关于 Git 的基础和进阶的知识，让读者在学习前端框架之余，还能学习到主流的团队代码管理工具和版本控制工具 Git 的知识应用。

本书共 11 章，分为基础篇、进阶篇和项目实战篇。基础篇（第 1～3 章）详细讲述 Vue.js 框架的基础知识点，并以 HTML 文件的方式切入，浅显易懂，让读者的学习体验达到最优。进阶篇（第 4～8 章）从 Webpack 起步，系统地讲述 Webpack 项目搭建、脚手架来源并切入基于脚手架的开发模式中。在进阶篇中还讲到 Vue.js 的高级语法（包括 Vuex 和 Vue 路由）的应用及 Vue.js 3.0 新增的语法。项目实战篇（第 9～11 章），分别用 Vue.js 2.0 技术实现了通用后台管理系统和大型 PC 商城两个实战项目。本书示例代码丰富，实用性和系统性较强，并配有视频讲解，助力读者透彻理解书中的重点、难点。

本书适合有少量 HTML+CSS+JavaScript 基础的初学者入门，并可作为高等院校和培训机构相关专业的教学参考书。

本书封面贴有清华大学出版社防伪标签，无标签者不得销售。
版权所有，侵权必究。举报：010-62782989，beiqinquan@tup.tsinghua.edu.cn。

**图书在版编目（CIP）数据**

Vue.js光速入门及企业项目开发实战/庄庆乐，任小龙，陈世云编著. —北京：清华大学出版社，2024.2
ISBN 978-7-302-65448-3

Ⅰ. ①V… Ⅱ. ①庄… ②任… ③陈… Ⅲ. ①网页制作工具－程序设计 Ⅳ. ①TP393.092.2

中国国家版本馆CIP数据核字（2024）第042630号

**责任编辑：**赵佳霓
**封面设计：**刘　键
**责任校对：**郝美丽
**责任印制：**刘海龙

**出版发行：**清华大学出版社
　　　　**网　　址：**https://www.tup.com.cn，https://www.wqxuetang.com
　　　　**地　　址：**北京清华大学学研大厦A座　　**邮　编：**100084
　　　　**社 总 机：**010-83470000　　**邮　购：**010-62786544
　　　　**投稿与读者服务：**010-62776969，c-service@tup.tsinghua.edu.cn
　　　　**质 量 反 馈：**010-62772015，zhiliang@tup.tsinghua.edu.cn
　　　　**课 件 下 载：**https://www.tup.com.cn，010-83470236
**印 装 者：**北京同文印刷有限责任公司
**经　　销：**全国新华书店
**开　　本：**186mm×240mm　　**印　张：**17.75　　**字　数：**402千字
**版　　次：**2024年3月第1版　　**印　次：**2024年3月第1次印刷
**印　　数：**1～2000
**定　　价：**69.00元

产品编号：100071-01

# 前言
PREFACE

当前的 Web 开发正处于一个迅速发展和不断演变的时代，在这个过程中，Vue.js 作为一款优秀的前端框架，迅速地在业界赢得了广泛的认可和普及，目前国内市场还是以 Vue.js 2.0 为主导。作为 Vue.js 的主要版本，Vue.js 2.0 不仅继承了 Vue.js 的核心概念和优势，还带来了更多的改进和增强，为我们构建强大、灵活、高效的 Web 应用提供了更多的可能。

笔者最早接触的前端框架技术是 jQuery 技术，随着 2013 年 React 的面世，到 2014 年引入中国，笔者开始接触 React 项目，成为中国第一批"吃螃蟹"的人。目前 Vue.js 2.0 的盛行和 Vue.js 3.0 的来临，笔者意识到 Vue.js 2.0 是前端初学者最好入手且是目前国内前端市场应用最广的前端框架，所以本书并没有从 Vue.js 3.0 开始介绍。笔者认为这个学习过程是逐层递进的，对初学者打好坚实的框架基础非常有帮助。此外，笔者还精心录制了一套前端的学习视频（前端基础到进阶 jQuery、Vue.js 2.0、微信小程序、React、Vue.js 3.0、TypeScript 等），目前播放量已近八十万人次。

笔者写作本书的目的是想传播前端框架领域知识（因为 Vue.js 是国人所编写，并且确实优秀），想为 Web 前端社区做一些贡献的同时也为叮丁狼教育的产品打下坚实的技术基础。写作本书期间，笔者查阅了大量的资料，使知识体系扩大了不少，收获良多。

## 本书主要内容

第 1 章介绍 Vue.js 框架、安装方式及其基础语法结构。

第 2 章介绍常用及不常用的指令（包括事件绑定指令及属性绑定指令等），以及计算属性。

第 3 章介绍过滤器、组件化开发及书店购物车项目实战。

第 4 章介绍 Webpack 及 Vue.js 的官方脚手架 Vue CLI 等。

第 5 章介绍 Vue.js 高级语法，如 Vue 插槽、修饰符、监听、动态组件和缓存及其他高级语法。

第 6 章介绍 DevTools 开发工具的使用及 Vuex 的核心概念等。

第 7 章介绍 Vue Router 及后端数据请求。

第 8 章介绍 Vue.js 3.0 新增语法。

第 9 章介绍 Vue.js 2.0 全家桶+Element 开发后台管理系统项目，该项目可在企业后台管理项目中直接套入使用。

第 10 章介绍 Git 基础知识和进阶使用、Git Flow 工作流模型，以及拓展了 Git 高级用法。

第 11 章介绍大型 PC 商城项目实战，并提供了用户行为验证、微信扫码登录、购物车、触底加载等常见的解决方案。

## 阅读建议

本书是一本基础入门加实战的书籍，既有基础知识，又有丰富示例，包括详细的操作步骤，实操性强。

关于学习的方式，建议读者从第 1 章开始（不建议零基础读者跳看 Vue.js 3.0 的章节），阅读完每个知识点案例之后，根据书中（或者资料中）给到的代码输入一遍，并理解，这个过程还要结合书中给的代码进行阅读理解，有助于更快速地掌握和理解知识点。

从第 4 章开始，进入 Webpack 的学习，这一章可以快速阅读，如遇问题不用在相关问题中纠结太久，反而会影响学习节奏。可以尝试先学习 Vue CLI 的安装和使用，并往后学习。因为做项目实战，用的就是 Vue CLI，所以直接上手亦可。待后期做完项目后再返回学习并深究 Webpack。

第 6 章和第 7 章是做项目时会运用到的重点，写到 Vuex 和 Vue Router，建议结合书中的视频来学习，这样会更加容易理解和上手。

第 9 章和第 11 章属于项目实战部分，读者在掌握了前面的基础知识后，可以通过两个实战项目学习企业应用的一些解决方案。在学习过程中遇到 Bug 不必紧张，程序员开发项目遇到 Bug 是正常的，通过百度、谷歌甚至 ChatGPT 翻译一下，即可得到解决方案。

扫描目录上方二维码可下载本书源码。

## 致谢

感谢单位领导及同事。特别感谢前端团队的老师们，帮助我处理好团队中的其他事务，使我可以全身心地投入写作工作之中。

由于时间仓促，书中难免存在不妥之处，请读者见谅，并提宝贵意见。

庄庆乐

2023 年 12 月

# 目 录
CONTENTS

本书源码

## 第 1 章 Vue.js 基础语法 ··· 1

### 1.1 Vue.js 框架简介 ··· 1
1.1.1 国内外前端主流框架分析 ··· 1
1.1.2 Vue.js 框架简介 ··· 2
1.1.3 Vue.js 开发编辑器 ··· 2
1.1.4 Vue.js 安装 ··· 3

### 1.2 Vue.js 语法结构 ··· 4
1.2.1 实例化 Vue 与 Mustache 语法 ··· 4
1.2.2 双向数据绑定及其原理 ··· 6
1.2.3 Vue.js 文件中的 MVVM ··· 8

## 第 2 章 Vue.js 指令、事件与计算属性 ··· 9

### 2.1 Vue.js 基础指令 ··· 9
### 2.2 事件绑定指令 ··· 14
### 2.3 属性绑定指令 ··· 17
### 2.4 计算属性 ··· 21

## 第 3 章 过滤器及组件化开发 ··· 25

### 3.1 过滤器与生命周期 ··· 25
3.1.1 Filter 过滤器 ··· 25
3.1.2 LifeCycle 生命周期 ··· 27

### 3.2 组件化开发 ··· 28
3.2.1 组件化开发的必要性 ··· 28
3.2.2 全局组件 ··· 28
3.2.3 局部组件 ··· 33

### 3.3 书店购物车项目实战 ··· 39

## 第 4 章 Webpack、Slot 与 Vue CLI 脚手架 — 46

### 4.1 Webpack 模块化打包工具 — 46
- 4.1.1 Webpack 的简介与安装 — 46
- 4.1.2 Webpack 基本配置 — 48
- 4.1.3 webpack-dev-server — 49
- 4.1.4 html-webpack-plugin — 50
- 4.1.5 loader — 51
- 4.1.6 babel — 52
- 4.1.7 HTML 热更新 — 53
- 4.1.8 图片资源 — 54

### 4.2 Vue CLI — 55
- 4.2.1 Vue CLI 的简介与安装 — 55
- 4.2.2 Vue CLI 创建项目 — 56
- 4.2.3 Vue CLI 项目预览 — 58

## 第 5 章 Vue.js 高级语法 — 61

### 5.1 插槽 — 61
- 5.1.1 匿名插槽 — 61
- 5.1.2 具名插槽 — 62
- 5.1.3 作用域插槽 — 63

### 5.2 修饰符 — 64
- 5.2.1 表单修饰符 — 64
- 5.2.2 事件修饰符 — 65
- 5.2.3 按键修饰符 — 66

### 5.3 监听 — 66
- 5.3.1 普通监听 — 66
- 5.3.2 立即监听 — 67
- 5.3.3 深度监听 — 68
- 5.3.4 deep 优化 — 69

### 5.4 动态组件与组件缓存 — 70
- 5.4.1 动态组件 — 70
- 5.4.2 KeepAlive 缓存组件 — 71

### 5.5 Vue.js 其他高级用法 — 72

## 第 6 章 Vuex — 76

### 6.1 DevTools — 76

6.2 Vuex ······················································································································· 79
　　6.2.1　Vuex 简介与安装 ········································································································ 79
　　6.2.2　Vuex 核心概念 ············································································································ 80

# 第 7 章　路由与请求 ·············································································································· 85

7.1 路由 ······················································································································· 85
　　7.1.1　Vue Router 简介与安装 ································································································ 85
　　7.1.2　路由文件配置 ·············································································································· 86
　　7.1.3　路由跳转 ···················································································································· 87
　　7.1.4　导航守卫 ···················································································································· 88
7.2 请求 ······················································································································· 90

# 第 8 章　Vue.js 3.0 新增语法 ································································································· 93

8.1 Vue.js 3.0 起步 ········································································································· 93
8.2 Vue.js 3.0 新增语法 ··································································································· 94
　　8.2.1　Composition API ········································································································· 94
　　8.2.2　Provide 与 Inject ········································································································ 98
　　8.2.3　Teleport ···················································································································· 98
　　8.2.4　Suspense ··················································································································· 99
　　8.2.5　Fragment ················································································································· 100
　　8.2.6　TreeShaking ············································································································· 101
　　8.2.7　Performance 提升 ······································································································ 102
　　8.2.8　生命周期 ·················································································································· 102

# 第 9 章　项目一：Vue.js 2.0 全家桶+Element 开发后台管理系统 ············································ 103

9.1 创建项目与添加 Element 模块 ···················································································· 103
9.2 项目初始化 ············································································································ 104
9.3 登录组件的初步引入及使用 ······················································································· 105
9.4 登录组件的初步完善 ································································································ 106
　　9.4.1　登录页面 ·················································································································· 106
　　9.4.2　覆盖 Element UI 样式的正确写法 ················································································· 108
　　9.4.3　书写校验规则 ············································································································ 108
　　9.4.4　自定义校验规则 ········································································································· 109
　　9.4.5　校验 ························································································································· 109
　　9.4.6　企业级项目验证 ········································································································· 110
　　9.4.7　验证码图片的获取 ····································································································· 111

9.5 封装 axios 的拦截器 ... 111
9.6 完善登录模块 ... 112
9.7 错误提示及其统一处理方案 ... 115
9.8 登录成功后跳转到首页 ... 117
9.9 经典三栏布局解决方案 ... 118
9.10 书写路由守卫 ... 120
9.11 手写菜单栏 ... 121
 9.11.1 折叠"菜单"按钮的初步规划 ... 121
 9.11.2 菜单展开和折叠状态的展示 ... 122
 9.11.3 是否折叠导航栏 ... 123
 9.11.4 修改 Vuex 中 isNavCollapse 的值 ... 124
 9.11.5 菜单栏折叠卡顿的问题 ... 125
 9.11.6 折叠过渡效果的实现 ... 125
 9.11.7 补充 Logo 和标题 ... 126
 9.11.8 定义初始数据导航 ... 126
 9.11.9 菜单实现路由跳转 ... 128
9.12 统一处理请求后的 code==200 的情况 ... 129
9.13 动态生成菜单栏 ... 131
 9.13.1 请求获取用户菜单列表 ... 131
 9.13.2 分析思路 ... 134
 9.13.3 处理 menuData 数组 ... 135
9.14 修改二级菜单栏的样式补充 ... 137
9.15 图标处理 ... 138
9.16 认证失败处理 ... 139
9.17 配置子路由(内容部分) ... 139
9.18 动态添加子路由规则 ... 140
9.19 添加路由切换的过渡动画 ... 142
9.20 面包屑处理 ... 143
 9.20.1 渲染和样式初步处理 ... 143
 9.20.2 title 的收集 ... 144
 9.20.3 在面包屑组件中展示 title ... 145
 9.20.4 解决网址栏跳转但视图不更新的情况 ... 145
9.21 404 页面的处理 ... 146
9.22 删除 token ... 147
9.23 用户信息处理 ... 147
 9.23.1 登录成功获取用户信息 ... 147

|     |        | 9.23.2 下拉菜单及退出登录 | 150 |
| --- | --- | --- | --- |
|     | 9.24   | 标签栏处理 | 151 |
|     |        | 9.24.1 初步布局 | 151 |
|     |        | 9.24.2 组织 tags 数组 | 152 |
|     |        | 9.24.3 当前样式的处理 | 153 |
|     |        | 9.24.4 跳转处理 | 155 |
|     |        | 9.24.5 删除标签 | 155 |
|     |        | 9.24.6 右击出现快捷菜单 | 156 |
|     |        | 9.24.7 菜单项现实逻辑的控制 | 156 |
|     |        | 9.24.8 静动态路由的区分 | 158 |
|     |        | 9.24.9 关闭标签栏 | 159 |
|     |        | 9.24.10 根据单击的项目对 tags 进行操作 | 160 |
|     | 9.25   | 表格处理 | 161 |
|     | 9.26   | 分页处理 | 164 |
|     | 9.27   | 导出文件与上传文件的处理 | 165 |
|     |        | 9.27.1 导出文件 | 165 |
|     |        | 9.27.2 上传文件 | 166 |

## 第 10 章 Git 介绍 ......169

| 10.1 | Git 的基本使用 | 170 |
| --- | --- | --- |
| 10.2 | Git Flow 工作流模型 | 174 |
| 10.3 | Git 拓展 | 176 |

## 第 11 章 项目二：大型 PC 商城 ......182

| 11.1 | 项目准备 | 182 |
| --- | --- | --- |
| 11.2 | 网站数据请求模块 | 183 |
| 11.3 | 头部组件 | 186 |
|      | 11.3.1 版心样式 | 186 |
|      | 11.3.2 头部组件布局 | 186 |
| 11.4 | 导航组件 | 188 |
|      | 11.4.1 基本布局 | 188 |
|      | 11.4.2 搜索框布局 | 189 |
|      | 11.4.3 路由配置及导航项当前样式 | 190 |
| 11.5 | 登录模块布局 | 192 |
|      | 11.5.1 模态窗口的书写 | 192 |
|      | 11.5.2 设置单击展示模态窗口 | 193 |

| | | |
|---|---|---|
| 11.5.3 | 单击关闭模态窗口 | 195 |
| 11.5.4 | 单击标题栏的切换效果 | 196 |
| 11.5.5 | 表单基本布局 | 197 |
| 11.6 | 拼图验证滑块 | 199 |
| 11.7 | 单击"获取验证码"按钮的逻辑 | 201 |
| 11.7.1 | 逻辑分析 | 201 |
| 11.7.2 | 判断手机号格式 | 201 |
| 11.7.3 | 倒计时及其展示 | 202 |
| 11.7.4 | 连续单击倒计时 Bug | 203 |
| 11.7.5 | 抽取工具函数 | 204 |
| 11.7.6 | 发起获取验证码请求 | 205 |
| 11.7.7 | 请求成功回调函数的完善 | 206 |
| 11.8 | 手机号码登录逻辑分析 | 207 |
| 11.8.1 | 抽取前两个验证的代码 | 207 |
| 11.8.2 | 发起登录请求 | 208 |
| 11.8.3 | 登录成功后的逻辑 | 209 |
| 11.8.4 | 购物车按钮的布局 | 210 |
| 11.8.5 | 购物车按钮展示（登录状态）分析 | 211 |
| 11.9 | 提示组件的封装 | 212 |
| 11.9.1 | icon 图标的使用 | 212 |
| 11.9.2 | Toast 组件的初步封装与使用 | 213 |
| 11.9.3 | Toast 组件展示 | 215 |
| 11.9.4 | Toast 组件的进场离场效果 | 216 |
| 11.9.5 | 封装 Toast 的属性 | 217 |
| 11.9.6 | Toast 组件自动关闭的处理 | 218 |
| 11.9.7 | 总结：提示框组件的使用 | 219 |
| 11.10 | 微信扫码登录——微信登录二维码的获取与展示 | 220 |
| 11.10.1 | 获取微信二维码 | 220 |
| 11.10.2 | 微信二维码样式调整 | 220 |
| 11.11 | 微信扫码登录——用临时票据 code 换取 token | 221 |
| 11.12 | 手机验证码登录 | 223 |
| 11.13 | 路由监听及其应用 | 223 |
| 11.14 | 组件重载 | 225 |
| 11.15 | 获取登录用户信息 | 226 |
| 11.16 | 用户信息渲染 | 227 |
| 11.17 | 删除 token 后的用户信息初始化 | 230 |

| | | |
|---|---|---|
| 11.18 | 首页布局的套用 | 230 |
| 11.19 | 详情页的处理 | 240 |
| 11.20 | 单击加入购物车 | 247 |
| 11.21 | 全部商品页面 | 248 |
| | 11.21.1 结构样式套用 | 248 |
| | 11.21.2 商品列表渲染 | 250 |
| | 11.21.3 选项数据的分析和渲染 | 251 |
| | 11.21.4 单击选项，切换商品列表 | 252 |
| | 11.21.5 搜索框事件 | 253 |
| 11.22 | 导航守卫 | 254 |
| | 11.22.1 全局导航守卫 | 254 |
| | 11.22.2 组件内部导航守卫 | 255 |
| 11.23 | 个人中心——购物车页面 | 255 |
| 11.24 | 404 处理 | 266 |
| 11.25 | 滚动到底部加载更多 | 266 |
| 11.26 | 跨域配置 | 269 |
| 11.27 | 项目环境变量配置 | 270 |

# 第 1 章 Vue.js 基础语法

## 1.1 Vue.js 框架简介

### 1.1.1 国内外前端主流框架分析

前端是个高度依赖框架实现快速开发的工作，对于很多前端初学者来讲并不知道各种主流框架之间的区别，本节主要阐述前端框架之间的区别与国内外前端主流框架的分析。

#### 1. 前端常用框架类型分析

类似于国内比较火的 Bootstrap、VantUI 与 ElementUI 这类框架，主要用于实现快速页面布局，框架设计团队会将一些常用的前端组件封装好，使框架使用者能够快速地完成组件调用，而不需要每个项目都实现一次类似于焦点图、穿梭框或面包屑等功能，这样可以极大地提升前端开发效率。

还有一些框架，如 Axios、Fetch 等框架，则主要用来对前端 HTTP 请求进行一系列封装，简化前端页面请求的代码，这些框架目前大多是基于 ES6 的 Promise 去实现的，在前端开发中，二者也经常被引入项目框架。

而类似于 jQuery、Vue.js 和 React.js 等框架，主要是用来实现对页面的非静态开发，其中 jQuery 与后两者完全不同，jQuery 是基于 DOM 操作的框架，主要提高事件操作上的体验，而 Vue.js 与 React.js 这类框架，更多地偏向于处理数据，通过数据驱动视图更新的形式，结合虚拟 DOM 实现对页面的一系列操作。

至于 Three.js、Node.js、TypeScript 等前端流行框架都有其各自的功能和特殊的应用场景，在此就不做过多赘述。

#### 2. 国外前端主流框架

在市场的占有率上，目前全世界运行中的项目，jQuery 仍稳居榜首，毕竟类似于 React 这种 MVC 框架流行起来的时间相对于 jQuery 较短，而 jQuery 在传统项目中早已占据了优势，但随着 MV*（包含 MVC、MVP 和 MVVM）框架的崛起，jQuery 让出宝座也只是时间问题。

目前国外相对流行的框架主要有 React.js 和 Angular.js 等，其中 React.js 起源于脸书（这里补充一下，如今脸书已正式更名为 Meta，下文将以 Meta 称呼该公司）的内部项目，用来架设 Instagram 网站，并于 2013 年 5 月开源。React.js 目前在国内也相对较火，这主要源于其社区的开放性，而 Angular.js（当前已被谷歌收购）由于更新迭代太快，国内早期用得比较多，至今很多项目也已经弃用，使用 Vue.js 或 React.js 进行了重构。

#### 3. 国内前端主流框架

从上文可以得知，React.js 和 Angular.js 在国内均有其一席之地，实际使用率上 React.js 更多一些，然而，二者有个共同特征：社区基本以英文居多。这对世界上绝大部分的前端工程师肯定是好事，但对于国内的前端开发人员则并不友好。另外，React.js 的上手难度较高，因此初学者往往很难简单地通过文档去掌握它，绝大部分的 React.js 项目存在于国内大厂，而中小企业或项目则更倾向于 Vue。在这个背景下，Vue 成为国内目前最为流行的前端 MVVM 框架。

### 1.1.2 Vue.js 框架简介

#### 1. Vue.js 官方简介

Vue（读音 /vju:/，类似于 view）是一套用于构建用户界面的渐进式框架。与其他大型框架不同的是，Vue.js 被设计为可以自底向上逐层应用。Vue.js 的核心库只关注视图层，不仅易于上手，还便于与第三方库或既有项目整合。另一方面，当与现代化的工具链及各种支持类库结合使用时，Vue.js 也完全能够为复杂的单页应用提供驱动。

#### 2. Vue.js 框架作者

根据百度百科的描述：尤雨溪，前端框架 Vue.js 的作者，HTML5 版 Clear 的打造人，独立开源开发者。曾就职于 Google Creative Labs 和 Meteor Development Group。由于工作中大量接触开源的 JavaScript 项目，最后自己也走上了开源之路，现全职开发和维护 Vue.js。

#### 3. Vue.js 框架现状

笔者从事前端开发与前端培训多年，对前端的就业行情有着相对较可观的了解。国内前端开发人员很大一部分是通过 IT 培训输出到市场上的，其中这部分学员寻找 Vue.js 的相关工作占比超过 90%，并且培训机构及高等院校大多也会将 Vue.js 列为前端体系的主要课程。在这种前提下，Vue.js 实现了弯道超车，在这个前端框架遍地开花的时代，后来者居上，成为国内前端最流行的框架之一。

同时，不得不承认，Vue.js 在一定程度上也参照了 React.js 等优秀框架的设计理念，而且文档与社区更加适合初学者上手，用法也相对简单。

### 1.1.3 Vue.js 开发编辑器

前端开发的编辑器有 WebStorm、VS Code 等神器，但真正意义上做到容易使用、插件多、兼容性强且免费的，这里推荐 VS Code。

VS Code 的全称为 Visual Studio Code，可通过 VS Code 官网直接下载，如图 1-1 所示。该软件是免费开源的，本书后续的课程均是通过 VS Code 进行开发。

图 1-1　Visual Studio Code 官网

本节主要阐述前端主流框架的分析、Vue.js 框架的简介与作者，以及开发编辑器。

### 1.1.4　Vue.js 安装

学习一个框架，需要参考其官网 https://cn.vuejs.org/，官网界面如图 1-2 所示。

图 1-2　Vue 官网

通过首页的"起步"按钮可以进入教程文档，单击"安装"按钮即可了解 Vue.js 的安装方式。

#### 1. CDN 引入安装

如果开发普通的 HTML 页面，想要体验 Vue.js，则可以使用 CDN 的形式引入 Vue 框架，在 HTML 的 head 标签中加入链接引用代码，代码如下：

```
<!-- 开发环境版本,包含了有帮助的命令行警告 -->
<script src="https://cdn.jsdelivr.net/npm/vue/dist/Vue.js"></script>
<!-- 生产环境版本,优化了尺寸和速度 -->
<script src="https://cdn.jsdelivr.net/npm/vue"></script>
```

这里建议初学者直接把 src 中的 URL 粘贴到浏览器中,全选后复制代码,在项目中新建 Vue.js 文件,并粘贴刚刚复制的代码。这样可以在无网状态下使用 VueCDN。

### 2. NPM 包管理器安装

后续通过 Vue CLI 5（Vue.js 官方的脚手架）方式引入,可以使用该方式,命令如下:

```
#最新稳定版
$ npm install vue
```

本节主要讲解了 Vue 的引入方式。在后续的章节中,会优先使用 CDN,直到讲述 Webpack 之后,才会开始介绍 Vue CLI 的具体项目创建方式。

## 1.2 Vue.js 语法结构

### 1.2.1 实例化 Vue 与 Mustache 语法

在原生 JavaScript 中,使用 new 可以实例化一个对象,初步体验 Vue.js 的语法可知,Vue.js 也采用相同的做法。

【示例 1-1】在页面的 h1 标签中显示"你好,世界",但这串文字不能将固定值写在标签中,要求使用 Vue.js 进行管理。

代码如下:

```
<!-- 第1章 Vue.js 基础语法: Mustache 语法使用方式 -->
<!DOCTYPE html>
<html lang="en">
<head>
    <meta charset="UTF-8">
    <meta name="viewport" content="width=device-width, initial-scale=1.0">
    <title>Document</title>
<script src="./Vue.js"></script>
</head>
<body>
    <div id="app">
        {{msg}}
    </div>
</body>
</html>
<script>
    //创建 Vue.js 对象
    new Vue({
```

```
        el: "#app",//指明挂载的元素 ID
        data: {//定义数据
            msg: "你好，世界"
        }
    })
</script>
```

上面的代码做了什么事情？

（1）创建了一个 Vue.js 对象。

（2）在创建 Vue 对象时，传入了一个 options 对象，options 中的 el 属性决定了这个 Vue 对象挂载到哪一个元素上，很明显，这里挂载到了 id 为 app 的元素上。

（3）同时 options 对象中包含了 data 属性，该属性中通常会存储一些数据，上面例子中的 msg 就是直接定义出来的数据。

（4）标签中使用{{}}包裹着 msg，这对双花括号就是 Mustache 语法，俗称胡子语法。

当然，data 中的数据除了可以像 msg 一样是一个字符串外，也可以是一个对象，因此，渲染方式也应当进行修改，代码如下：

```
<!-- 第 1 章 Vue 基础语法：Vue 基本语法 -->
<!DOCTYPE html>
<html lang="en">
<head>
    <meta charset="UTF-8">
    <meta name="viewport" content="width=device-width, initial-scale=1.0">
    <title>Document</title>
<script src="./Vue.js"></script>
</head>
<body>
    <div id="app">
        姓名：{{obj.name}}
        <hr />
        年龄：{{obj.age}}
    </div>
</body>
</html>
<script>
    //实例化
    new Vue({
        el: "#app",
        data: {
            obj: {
                name: "张三",
                age: 34
            }
        }
    })
```

```
</script>
```

以上代码主要运用了 Vue.js 的基本语法规则。

## 1.2.2 双向数据绑定及其原理

### 1. Vue.js 双向数据绑定场景

【**示例 1-2**】 页面中有一个输入框 input 和一个标题标签 h1，当 input 输入内容时，想看到 h1 的内容也被同步修改了。

代码如下：

```
<!-- 第1章 Vue.js 基础语法：Vue 双向数据绑定 -->
<!DOCTYPE html>
<html>
<head>
    <meta charset="UTF-8">
    <meta name="viewport" content="width=device-width, initial-scale=1.0">
    <meta http-equiv="X-UA-Compatible" content="ie=edge">
    <title></title>
</head>
<body>
<div id="app">
    <input type="text" v-model="word">
    <h2>{{word}}</h2>
</div>
</body>
</html>
<script src="./Vue.js"></script>
<script>
var vm = new Vue({
    el: "#app",
    data: {
        word: "你好世界"
    }
})
</script>
```

在以上代码中，使用 v-model 指令达到了这个效果。尽管读者暂时不清楚 v-model 的作用，但依然可以观察到 Vue.js 在处理数据方面，确实会让开发者感受到其便捷性。

### 2. 简单实现双向数据绑定

【**示例 1-3**】 接下来用原生 JavaScript 实现一个简单的双向数据绑定。

代码如下：

```
<!-- 第1章 Vue.js 基础语法：简单实现双向数据绑定 -->
<!DOCTYPE html>
<html lang="zh_cn">
```

```html
<head>
    <meta charset="UTF-8">
    <meta http-equiv="X-UA-Compatible" content="IE=edge">
    <meta name="viewport" content="width=device-width, initial-scale=1.0">
    <title>Document</title>
</head>
<body>
    <input type="text" id="ipt" value="" oninput="iptChange(this.value)" />
    <h3 id="title"></h3>
</body>
</html>
<script>
//获取元素
var ipt = document.getElementById('ipt');
var title = document.getElementById('title');

//声明一个data对象,专门用来存放数据
var data = {};

/*
    用来给对象动态添加属性,define 表示定义,Property 表示属性
    Object.defineProperty(对象的名称,属性的名称,配置项)
    get 代表定义这个msg时要给它赋值的数据
    set 代表当获得最新的值时,想要去做什么事
*/
Object.defineProperty(data, 'msg', {
    get: function(){
        return "你好世界"
    },
    set: function(newVal){
        //console.log(newVal)   //msg 已经被修改了
        title.innerHTML = newVal;
        ipt.value = newVal;
    }
})

//console.log(data.msg)
title.innerHTML = data.msg;
ipt.value = data.msg;

//input 值发生了改变
function iptChange(val){
    //修改数据源(data.msg),会自动触发set方法
    data.msg = val;
}
</script>
```

### 3. 双向数据绑定原理

Vue.js 双向数据绑定原理是通过数据劫持结合发布订阅者模式的方式实现的，其核心是 Object.defineProperty()方法。实现的基本效果是数据和视图同步，当数据发生变化时视图跟着变化、视图变化时数据也随之发生改变。

Object.defineProperty(obj, prop, descriptor)有 3 个参数，分别为 obj（要定义属性的对象）、prop（要定义或修改的属性）、descriptor（具体的改变方法）。简单地说，就是用这种方法来定义一个值。当调用时，使用了 Object.defineProperty()中的 get()方法，当给这个属性赋值时，又用到了 Object.defineProperty()中的 set()方法。

## 1.2.3 Vue.js 文件中的 MVVM

要理解 MVVM，可以进行如下拆分。

（1）View（V）层：视图层，在前端里指 DOM 层，主要作用是给用户展示各种信息。

（2）Model（M）层：数据层（逻辑层），数据可能是自定义的数据，或者从网络请求下来的数据。

（3）ViewModel（VM）层：视图模型层，是 View 层和 Model 层沟通的桥梁。一方面它实现了数据绑定（Data Binding），将 Model 的改变实时反映到 View 中；另一方面它实现了 DOM 监听，当 DOM 发生改变时可以相应地改变数据（Data）。

MVVM 的流程如图 1-3 所示。

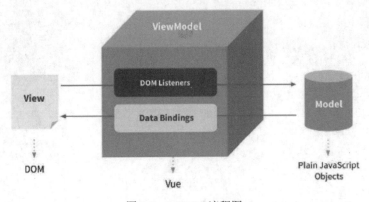

图 1-3　MVVM 流程图

本节主要讲解了 Vue.js 的基本语法结构、双向数据绑定及其原理，以及对 MVVM 进行了剖析。

# 第 2 章 Vue.js 指令、事件与计算属性

## 2.1 Vue.js 基础指令

学习 Vue 必然要了解这个框架的常用指令,它们可以大大提升前端开发人员的开发效率。

### 1. v-pre 跳过编译

当开发者希望在 HTML 中为用户显示出{{}}时,就需要用 v-pre 来通知 Vue.js 跳过当前这个标签的编译,代码如下:

```
<div v-pre>{{123}}</div>
```

以上代码可以让网页直接显示{{123}},而非 123。

### 2. v-html 标签解析

除了可以使用 Mustache 语法渲染文本,Vue.js 也可以使用 v-html 来渲染,同时要强调一下,v-html 不仅可以渲染文本,还可以解析标签,代码如下:

```
<!-- 第 2 章 Vue.js 指令、事件与计算属性:v-html 指令实现标签解析 -->
<!DOCTYPE html>
<html lang="en">
<head>
  <meta charset="UTF-8">
  <meta name="viewport" content="width=device-width, initial-scale=1.0">
  <meta http-equiv="X-UA-Compatible" content="ie=edge">
  <title>Document</title>
</head>
<body>
  <div id="app">
    <!--通过 v-html 指令将 htmlTxt 解析到 div 中-->
    <div v-html="htmlTxt"></div>
  </div>
</body>
<script src="https://cdn.jsdelivr.net/npm/vue/dist/Vue.js"></script>
<script>
  let vm = new Vue({
    el: '#app',
```

```
    data: {
      htmlTxt: '<p><strong>你好世界</strong></p>'
    }
  })
</script>
</html>
```

以上代码会将标签一起解析出来，打开网页可以看到文字是加粗的。

### 3. v-text 文本解析

从初学者的角度而言，基本可以认为 v-text 和 Mustache 语法的作用相同。v-text 与 v-html 最大的不同在于：v-html 可以解析 html，而 v-text 只能原封不动地把标签显示在页面中，无法解析成标签对应的格式，代码如下：

```
<!-- 第 2 章 Vue.js 指令、事件与计算属性：v-text 实现文本解析 -->
<!DOCTYPE html>
<html lang="en">
<head>
  <meta charset="UTF-8">
  <meta name="viewport" content="width=device-width, initial-scale=1.0">
  <meta http-equiv="X-UA-Compatible" content="ie=edge">
  <title>Document</title>
</head>
<body>
  <div id="app">
      <!--通过 v-text 指令将 htmlTxt 解析到 div 中-->
      <div v-text="htmlTxt"></div>
  </div>
</body>
<script src="https://cdn.jsdelivr.net/npm/vue/dist/Vue.js"></script>
<script>
  let vm = new Vue({
    el: '#app',
    data: {
      htmlTxt: '<p><strong>你好世界</strong></p>'
    }
  })
</script>
</html>
```

以上代码最终显示出来的将是<p><strong>你好世界</strong></p>。明显可以看出，无法解析标签。

### 4. v-cloak 斗篷

当使用{{}}这样的 Mustache 语法来将数据渲染到 HTML 中时，由于网络请求等原因造成卡顿，偶尔会导致用户先看到{{}}符号，过一会才能显示出来对应的数据。如果想要避免这种样式出现，则可以使用 v-cloak，代码如下：

```
<!-- 第2章 Vue.js 指令、事件与计算属性：v-cloak 斗篷 -->
<!DOCTYPE html>
<html lang="en">
<head>
  <meta charset="UTF-8">
  <meta name="viewport" content="width=device-width, initial-scale=1.0">
  <meta http-equiv="X-UA-Compatible" content="ie=edge">
  <title>Document</title>
  <style>
    [v-cloak] {
      display: none;
    }
  </style>
</head>
<body>
  <div id="app">
    <!--给div添加v-cloak指令-->
    <div v-cloak>hello {{textTxt}}</div>
  </div>
</body>
<script src="https://cdn.jsdelivr.net/npm/vue/dist/Vue.js"></script>
<script>
  let vm = new Vue({
    el: '#app',
    data: {
      textTxt: 'Vue'
    }
  })
</script>
</html>
```

将这份代码在浏览器打开，打开后在浏览器控制台中找到 Network，如图 2-1 所示，将其中的 No throttling 改为 slow 3G，这样就能看到效果了。

图 2-1　控制台 Network

### 5. v-show 显示

当开发者需要通过某个条件决定元素的显示或隐藏时，可以使用 v-show，这个属性的值可以直接填写表达式。当这个值的判断结果为 true 时，显示元素，否则无法显示元素，代码如下：

```html
<!-- 第2章 Vue.js指令、事件与计算属性：v-show 指令 -->
<!DOCTYPE html>
<html>
<head>
    <meta charset="UTF-8">
    <meta name="viewport" content="width=device-width, initial-scale=1.0">
    <meta http-equiv="X-UA-Compatible" content="ie=edge">
    <title></title>
</head>
<body>
<div id="app">
    <!--通过v-show控制满足条件的元素显示-->
    <h2 v-show="status == 1">1</h2>
    <h2 v-show="status == 2">2</h2>
    <h2 v-show="status == 3">3</h2>
</div>
</body>
</html>
<script src="./Vue.js"></script>
<script>
new Vue({
    el: "#app",
    data: {
        status: 1
    }
})
</script>
```

以上代码可以通过修改status的值，决定显示哪个h2标签。

### 6. v-if 条件判断与显示

v-if 也可以用来控制元素的显示与隐藏，代码如下：

```html
<!-- 第2章 Vue.js指令、事件与计算属性：v-if 指令 -->
<!DOCTYPE html>
<html>
<head>
    <meta charset="UTF-8">
    <meta name="viewport" content="width=device-width, initial-scale=1.0">
    <meta http-equiv="X-UA-Compatible" content="ie=edge">
    <title></title>
</head>
<body>
<div id="app">
    <!--通过v-if控制满足条件的元素渲染-->
    <h2 v-if="flag">你好世界</h2>
```

```
    </div>
  </body>
</html>
<script src="./Vue.js"></script>
<script>
new Vue({
    el: "#app",
    data: {
        //定义条件渲染的数据
        flag: true
    },
})
</script>
```

通过一个 flag，开发者就可以控制一个元素是否显示。那么 v-if 与 v-show 有什么区别呢？答案是：v-if 可根据条件渲染或者销毁元素，而 v-show 只会控制元素的显示和隐藏，相当于改变 display 值。

### 7. v-for 循环

Vue.js 文件中的循环使用 v-for 指令实现。假如现在有一个记录人物名称的数组，想要渲染在页面中，以列表的形式存在。实现的代码如下：

```
<!-- 第 2 章 Vue.js 指令、事件与计算属性：v-for 循环 -->
<ul>
    <!--通过 v-for 遍历数组，生成数组对应的数据，item 指当前遍历到数组数据，index 指数组下标-->
    <li v-for="(item,index) in arr">{{item}}</li>
<ul>
<script>
export default {
  data(){
      return {
        //定义列表渲染数据
        arr: ["刘备", "关羽", "张飞"]
      }
   }
}
</script>
```

其中，item 指数组中的每项，index 指每项的索引值。

### 8. v-bind:key 唯一标识符

实际上，在循环的过程中需要给元素的标签添加上 key 属性。如果没有 key 属性，当每次数据有局部更新时就需要整个数组重新渲染，对性能很不友好。给元素添加了 key 属性后，当数组有更改时，虚拟 DOM 会与真实 DOM 上的 key 值进行比较，如果真实 DOM 元素上的 key 值与虚拟 DOM 上的 key 值相同，则忽略该 DOM 元素的更新，从而提升渲染性能。

key 值作为唯一标识符，建议如果数组项中有 id，则优先考虑 id 作为 key 值。因为 id 是唯一的，当数组更新时，元素上的 key 就不会随之而变化。如果以 index 作为 key，当数组的增删导致元素重排时，key 值就会混乱，因此建议 key 值优先考虑 id。若已知数组是不会更新的，则可采用 index 等字段作为 key 值。

通过介绍 key 属性的作用，开发者需要知道，Vue 真正的 for 循环格式如下：

```
<!-- 第 2 章 Vue.js 指令、事件与计算属性：v-bind 指令 -->
<ul>
    <!-- 绑定 key -->
    <li v-for="item in arr" v-bind:key="item.id">{{item. username}}</li>
<ul>
<script>
export default {
  data(){
    return {
        //定义列表渲染的数组数据
        arr: [
            {id: 'm1', username: "刘备"},
            {id: 'm2', username: "关羽"},
            {id: 'm2', username: "张飞"},
        ]
    }
  }
}
</script>
```

**注意** v-bind:key 是给 key 属性动态绑定值的，目前还未讲述，第 3 章将会重点讲解。

Vue.js 的指令能为开发提升效率，是 Vue.js 学习路线中非常重要的一个概念。本节主要讲解了 Vue.js 的基础指令，也是部分常用的指令，如 v-html、v-show、v-if 和 v-for 等。后文还会继续讲述更多常用指令。

## 2.2 事件绑定指令

本节将介绍 Vue.js 的事件绑定指令，通过事件绑定指令可以实现页面的一系列操作。

### 1. v-on 绑定事件与 methods

使用 v-on 可以绑定一个事件，并且事件的名称需要在 methods 中定义，代码如下：

```
<!-- 第 2 章 Vue.js 指令、事件与计算属性：v-on 绑定事件与 methods 使用方法 -->
<!DOCTYPE html>
<html>
<head>
    <meta charset="UTF-8">
    <meta name="viewport" content="width=device-width, initial-scale=1.0">
```

```
    <meta http-equiv="X-UA-Compatible" content="ie=edge">
    <title></title>
</head>
<body>
<div id="app">
    <!-- 利用v-on:事件类型绑定事件 -->
    <button v-on:click="btnClick">按钮</button>
</div>
</body>
</html>
<script src="./Vue.js"></script>
<script>
    new Vue({
        el: "#app",
        methods: {
            //按钮单击事件, 目的是触发Chrome 控制台打印 123
            btnClick(){
            console.log(123);
            }
        }
    })
</script>
```

### 2. 事件操作 data 数据

既然 v-on 绑定的事件可以执行，就可以修改 data 中的数据，从而实现对逻辑层的数据更新。

【示例 2-1】 一个元素的显示与隐藏由 flag 控制，而单击按钮可以切换 flag 的值，从而实现对元素的显示与隐藏控制。

代码如下：

```
<!-- 第 2 章 Vue.js 指令、事件与计算属性：事件与 data -->
<!DOCTYPE html>
<html>
<head>
    <meta charset="UTF-8">
    <meta name="viewport" content="width=device-width, initial-scale=1.0">
    <meta http-equiv="X-UA-Compatible" content="ie=edge">
    <title></title>
</head>
<body>
<div id="app">
    <h2 v-show="flag">你好世界</h2>
    <button v-on:click="changeFlag">切换 flag</button>
</div>
</body>
</html>
```

```
<script src="./Vue.js"></script>
<script>
new Vue({
    el: "#app",
    data: {
        flag: true
    },
    methods: {
      changeFlag(){
          //对 data 中的 flag 进行赋值，值为 flag 自身的取反结果
          this.flag = !this.flag;
      }
    }
})
</script>
```

可以看到操控 data 中的数据 flag，其实就是操作 this 这个对象中的 flag 属性，写法为 this.flag。

### 3. 事件调用

当一个事件在执行的过程中需要调用其他已经封装好的事件时，可以使用 this.函数名() 的格式进行调用，代码如下：

```
<!-- 第 2 章 Vue.js 指令、事件与计算属性：事件调用 -->
<!DOCTYPE html>
<html>
<head>
    <meta charset="UTF-8">
    <meta name="viewport" content="width=device-width, initial-scale=1.0">
    <meta http-equiv="X-UA-Compatible" content="ie=edge">
    <title></title>
</head>
<body>
<div id="app">
<h2 v-show="flag">你好世界</h2>
<button v-on:click="changeFlag">切换 flag</button>
</div>
</body>
</html>
<script src="./Vue.js"></script>
<script>
new Vue({
    el: "#app",
    data: {
        flag: true
    },
    methods: {
```

```
        changeFlag(){
            //对 data 中的 flag 进行赋值,值为 flag 自身的取反结果
            this.flag = !this.flag;
            //除了取反,如果还想实现打印 123 之类的操作,则可以调用 logFn()方法来执行
            this.logFn();
        },
        logFn(){
            //打印 123
            console.log(123)
        }
    }
})
</script>
```

#### 4. v-on 的简写

在前文叙述中,已经对 v-on 指令的主要使用方式进行了讲解,而在实际的开发过程中,往往使用的都是 v-on 的简写形式@,代码如下:

```
<button v-on:click="changeFlag">切换 flag</button>
<!-- v-on 可以简写为@ -->
<button @click="changeFlag">切换 flag</button>
```

本节主要讲解了事件绑定指令 v-on 的用法、如何修改 data 中的数据、如何调用其他事件,以及 v-on 的简写。

## 2.3 属性绑定指令

本节继续介绍 Vue.js 的常用指令,并了解这部分指令之间的区别。

#### 1. v-bind 属性绑定

【示例 2-2】 一个 HTML 标签的属性,如果想要绑定动态值,就需要使用 v-bind 来绑定,代码如下:

```
<!-- 第 2 章 Vue.js 指令、事件与计算属性:v-bind 属性绑定 -->
<!DOCTYPE html>
<html>
<head>
    <meta charset="UTF-8">
    <meta name="viewport" content="width=device-width, initial-scale=1.0">
    <meta http-equiv="X-UA-Compatible" content="ie=edge">
    <title></title>
</head>
<body>
<div id="app">
<a href="https://baidu.com">跳转到百度</a>
<a v-bind:href="link">跳转到百度</a>
</div>
```

```
</body>
</html>
<script src="./Vue.js"></script>
<script>
new Vue({
    el: "#app",
    data: {
        link: "https://baidu.com"
    }
})
</script>
```

从以上代码可以看出，link 作为 data 中定义的字段，如果想要赋值给 a 标签上的 href 属性，就需要使用 v-bind 来绑定属性，这就是 Vue.js 文件中对属性进行绑定的指令。

注意　v-bind 与 v-on 一样，有其简写方式。v-bind 可以简写为冒号。通常 v-bind:href 可以写为:href，从而简化代码。从下文开始将使用 v-bind 的简写形式来描述属性绑定。

### 2. v-bind 绑定 class 属性

掌握了 v-bind 的用法，再来详细探讨针对 class 属性可以怎么绑定。之所以要进行探讨，原因在于 Vue.js 语法允许有多种表现形式。

**1）对象格式**

假设现在有若干类名，并已经写好了样式，代码如下：

```css
/* 控制文字及背景颜色 */
.fontblue{
    /* 将文字设置为蓝色 */
    color: blue;
}

.bgpink{
    /* 将背景设置为粉色 */
    background: pink;
}
```

让某个元素添加以上 fontblue 样式，或是两个样式都添加，代码如下：

```
<!-- 添加 fontblue 样式 -->
<li :class="{'fontblue': true}">蓝色文字</li>
<!-- 同时添加 fontblue 与 bgpink 两个样式 -->
<li :class="{'fontblue': true, 'bgpink': true}">蓝色文字,粉色背景</li>
```

在以上代码中 true 可以控制该元素享用其前方的类名，如果把 true 改为 false，则样式消失，因为类名将不再添加。可以看出，类名的添加与否取决于其键-值对的值。

**2）三元运算符格式**

对示例 2-2 进行变更：如果条件为 true，则让文字显示蓝色，如果条件为 false，则添加粉色背景。可以通过三元运算符实现，代码如下：

```
<li class="line" :class="true ? 'fontblue' : 'bgpink'">条件为true显示蓝色字体，条件为false显示粉色背景</li>
```

这里条件为true，所以文字将显示蓝色，而不显示粉色背景。

**3）数组格式**

再次对示例2-3进行变更，要求蓝色文字与粉色背景同时存在。此时除了对象的写法，还可以使用数组的形式来表现，代码如下：

```
<li :class="['fontblue', 'bgpink']">蓝色文字、粉色背景</li>
```

**4）函数格式**

在特殊情况下，开发者还可以使用函数调用的形式来返回类名，从而给元素添加样式，代码如下：

```
<li :class="getClassName()">蓝色文字、粉色背景</li>
```

另外，还需要在methods中定义该方法，并且返回样式字符串、样式对象或样式数组格式，代码如下：

```
new Vue({
    el: "#app",
    methods: {
        getClass(){
        //返回样式字符串格式
          return 'fontblue';
        //返回对象格式
          return {'fontblue': true};
        //返回数组格式
          return ['fontblue', 'bgpink'];
        }
    }
})
```

**3. v-bind 绑定 style 属性**

与class属性的绑定一样，style属性的绑定也有多种表现形式。在HTML中，假如有一个元素需要写行内式的样式（如实现"红色文字"），代码如下：

```
<div style="color: red;">标签文字</div>
```

**1）对象格式**

在Vue.js文件中，通过v-bind操作color的值，使其动态化，便可以任意控制文字的颜色，而不像以上代码，将color的值固定化。具体实现color值动态化的代码如下：

```
<!-- 此处:style相当于v-bind:style -->
<div :style="{color: fontColor}">标签文字</div>
```

```
<script>
    new Vue({
      data(){
          return {
              fontColor: "red"
          }
      }
    })
</script>
```

如果同时对多个样式进行控制,如同时实现红色文字、蓝色背景,则代码如下:

```
<!-- backgroundColor 也可以写为 'background-color' -->
<div :style="{color: fontColor, backgroundColor: bgColor}">标签文字</div>
<script>
    new Vue({
      data(){
          return {
              fontColor: "red",
              bgColor: "blue"
          }
      }
    })
</script>
```

**2)数组格式**

假如有多个样式对象,开发者希望集体呈现,那么还有数组格式,代码如下:

```
<div :style="[fontColor, bgColor]">标签文字</div>
<script>
    new Vue({
      data(){
          return {
              //每个字段代表一组样式
              fontColor: {color: 'red'},
              bgColor: {background: 'blue'}
          }
      }
    })
</script>
```

**3)函数格式**

通过函数返回值,也可以控制标签样式。这个返回值可以是对象或数组,代码如下:

```
<div :style="getStyle1()">获取对象格式样式</div>
<div :style="getStyle2()">获取数组格式样式</div>

<script>
```

```
    new Vue({
        data(){
            return {
                //每个字段代表一组样式
                fontColor: {color: 'red'},
                bgColor: {background: 'blue'}
            }
        },
        methods: {
            getStyles1(){
                return {color: 'red'}
            },
            getStyles2(){
                return [this.fontColor, this.bgColor]
            }
        }
    })
</script>
```

本节主要讲解了属性绑定的方式,并主要强调了 class 与 style 的绑定有多种形式,这需要在项目实战中具体了解它们的使用时机。

## 2.4 计算属性

本章将介绍 Vue 的计算属性,使用计算属性可以实现对数据的实时计算,为开发带来便捷。常见的使用场景为购物车模块开发等。

**1. computed 计算属性**

在实际开发中,往往会让用户在一个输入框中填写姓,再到另一个输入框填写名,此时想要实时得到用户的完整姓名,就可以借助计算属性,代码如下:

```
<!-- 第 2 章 Vue.js 指令、事件与计算属性:computed 计算属性 -->
<!DOCTYPE html>
<html>
<head>
    <meta charset="UTF-8">
    <meta name="viewport" content="width=device-width, initial-scale=1.0">
    <meta http-equiv="X-UA-Compatible" content="ie=edge">
    <title></title>
</head>
<body>
<div id="app">
    <div>
        <label for="xing">姓:</label>
        <input id="xing" v-model="xing" />
    </div>
```

```
        <div>
            <label for="ming">名:</label>
            <input id="ming" v-model="ming" />
        </div>
        <h3>用户姓名为{{username}}</h3>
</div>
</body>
</html>
<script src="./Vue.js"></script>
<script>
    new Vue({
        el: "#app",
        data: {
            return {
                xing: "",
                ming: ""
            }
        },
        //计算属性
        computed: {
            username(){
                return this.xing + ' ' + this.ming;
            }
        }
    })
</script>
```

> **注意** 在计算属性 computed 中定义过的字段,不能在 data 中重复定义。

### 2. computed 的 getter 与 setter

在 Vue.js 文件中定义每个属性都使用 Object.defineProperty()方法,该方法会为属性追加 get()方法与 set()方法,一般称为 getter 和 setter。

getter 主要用来返回数据结果,setter 的作用则是当该属性的值发生变化时,可以对外触发事件操作。

针对上述代码示例,可以使用 getter 与 setter 进行改写,代码如下:

```
<!-- 第2章 Vue.js指令、事件与计算属性: computed 的 getter 与 setter -->
<!DOCTYPE html>
<html>
<head>
    <meta charset="UTF-8">
    <meta name="viewport" content="width=device-width, initial-scale=1.0">
    <meta http-equiv="X-UA-Compatible" content="ie=edge">
    <title></title>
</head>
<body>
```

```html
<div id="app">
    <div>
        <label for="xing">姓：</label>
        <input id="xing" v-model="xing" />
    </div>
    <div>
        <label for="ming">名：</label>
        <input id="ming" v-model="ming" />
    </div>
    <div>
        <label for="username">用户姓名：</label>
        <input id="username" v-model="username" />
    </div>
</div>
</body>
</html>
<script src="./Vue.js"></script>
<script>
new Vue({
    el: "#app",
    data: {
        return {
            xing: "",
            ming: ""
        }
    },
    //计算属性
    computed: {
        username: {
            get(){//获取数据
                return this.xing + ' ' + this.ming;
            },
            set(val){//设置数据
                //val 为 username 的新值
                const u = val.split(' ');
                this.xing = u[0];
                this.ming = u[1];
            }
        }
    }
})
</script>
```

### 3. computed 与 methods 的区别

computed 可以计算一个值，methods 同样可以计算一个值，代码如下：

```
<!-- 第 2 章 Vue.js 指令、事件与计算属性：computed 与 methods 的区别 -->
```

```html
<!DOCTYPE html>
<html>
<head>
    <meta charset="UTF-8">
    <meta name="viewport" content="width=device-width, initial-scale=1.0">
    <meta http-equiv="X-UA-Compatible" content="ie=edge">
    <title></title>
</head>
<body>
<div id="app">
    <div>
      <label for="xing">姓: </label>
      <input id="xing" v-model="xing" />
    </div>
      <div>
        <label for="ming">名: </label>
        <input id="ming" v-model="ming" />
      </div>
    <h3>用户姓名为{{username}}</h3>
</div>
</body>
</html>
<script src="./Vue.js"></script>
<script>
new Vue({
    el: "#app",
  data: {
     return {
         xing: "",
         ming: ""
     }
  },
  //计算属性
  methods: {
     username(){
         return this.xing + ' ' + this.ming;
     }
  }
})
</script>
```

那么，computed 能完成的事情，methods 也能完成，为什么还需要 methods 呢？原因是 computed 有缓存，而 methods 没有，在一个实例中，当同一个 computed 属性被多次调用时只会执行一次方法。而 methods 中的方法，调用一次就执行一次，不存在缓存。从性能上来讲，同一个计算需求若 computed 与 methods 都能完成，则优先选择 computed。

本节主要讲解了 computed 计算属性，以及它的 getter 与 setter，同时也对比了 computed 与 methods。

# 第 3 章 过滤器及组件化开发

## 3.1 过滤器与生命周期

### 3.1.1 Filter 过滤器

在真实项目开发中，一些数据需要过滤出固定的格式再渲染，可以使用 Vue.js 的过滤器 Filter。Filter 分为全局过滤器与局部过滤器。

**1. 全局过滤器**

当一个页面有多个 Vue.js 实例时都想要共享一个过滤器，那么全局过滤器是首选，代码如下：

```html
<!-- 第3章过滤器及组件化开发：全局过滤器 -->
<!DOCTYPE html>
<html>
<head>
    <meta charset="UTF-8">
    <meta name="viewport" content="width=device-width, initial-scale=1.0">
    <meta http-equiv="X-UA-Compatible" content="ie=edge">
    <title></title>
</head>
<body>
<div id="app">
    <h3>过滤前的价格：{{price}}</h3>
    <!-- 竖线表示将前者的数据传入后者进行过滤 -->
    <h3>过滤后的价格：{{price | priceFilter}}</h3>
</div>
</body>
</html>
<script src="./Vue.js"></script>
<script>
//全局过滤器
Vue.filter('priceFilter', (val)=>{
    return "¥ " + Number(val).toFixed(2) + " 元";
})
```

```
new Vue({
el: "#app",
data: {
  price: 20.8
  }
})
</script>
```

过滤前的价格：20.8

过滤后的价格：￥20.80 元

图 3-1 价格过滤器

价格过滤器实现的效果如图 3-1 所示，价格从一个普通的数字，转变为项目中常用的价格显示格式。

2. 局部过滤器

如果一个过滤器只在当前组件（或实例）中使用，则局部过滤器也可以实现图 3-1 所示的效果，代码如下：

```
<!-- 第3章过滤器及组件化开发：局部过滤器 -->
<!DOCTYPE html>
<html>
<head>
    <meta charset="UTF-8">
    <meta name="viewport" content="width=device-width, initial-scale=1.0">
    <meta http-equiv="X-UA-Compatible" content="ie=edge">
    <title></title>
</head>
<body>
<div id="app">
    <h3>过滤前的价格：{ price}}</h3>
    <!--竖线表示将前者的数据传入后者进行过滤 -->
    <h3>过滤后的价格：{ price | priceFilter}}</h3>
</div>
</body>
</html>
<script src="./Vue.js"></script>
<script>
new Vue({
    el: "#app",
    data: {
        price: 20.8
    },
    //局部过滤器
    filters: {
        priceFilter(val){
            return "￥ " + Number(val).toFixed(2) + " 元";
        }
    }
})
</script>
```

## 3.1.2 LifeCycle 生命周期

每个程序从创建到销毁的过程被称为"生命周期"。Vue.js 的生命周期主要有 8 个，如图 3-2 所示，分别是 beforeCreate()、created()、beforeMount()、mounted()、beforeUpdate()、updated()、beforeDestroy()、destroyed()，这 8 个函数有时也称为钩子函数。

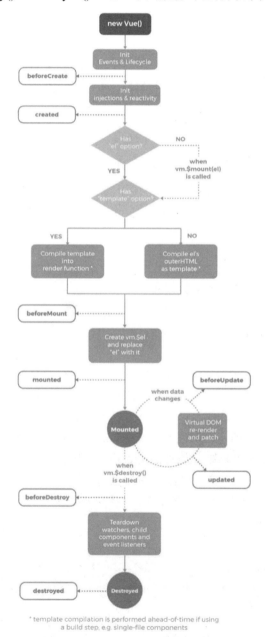

图 3-2　Vue.js 的生命周期

#### 1. 创建前后

一个实例（或组件）在创建前和创建后各由一个生命周期函数表示，分别是 beforeCreate() 和 created()。一般在 created() 函数中做页面初始数据的请求。

#### 2. 挂载前后

一个实例（或组件）渲染页面和数据的过程称为挂载。以挂载前与挂载后表示，对应的钩子函数分别为 beforeMount() 与 mounted()。一般来讲，到了 mounted() 这个生命周期，该实例（或组件）便已渲染完毕。

#### 3. 更新前后

一个实例（或组件）的数据通常会有更新，而更新也可分为更新前与更新后，其对应的钩子函数分别为 beforeUpdate() 与 updated()。

#### 4. 销毁前后

一个实例（或组件）在被销毁时往往需要做一些数据清除工作。销毁的生命周期分为销毁前与销毁后，分别是 beforeDestroy() 与 destroyed()。在这两个生命周期中可以对 Cookie、localStorage 所存放的数据进行必要的清理，或者手动对程序的垃圾进行回收等。

本节主要讲解了过滤器 Filter 的使用方法，以及生命周期 LifeCycle 各钩子函数的作用。

## 3.2 组件化开发

Vue.js 有两个最重要的核心概念，一是双向数据绑定，二是组件化开发，因此，本章所阐述的内容极为重要。

### 3.2.1 组件化开发的必要性

当一个带有数据、样式的 HTML 模块在一个项目中的多处被使用时，如果每个地方都重复写一遍，则会造成代码冗余。此时，可以考虑使用组件来解决这个问题。组件化开发在现在的网站开发中随处可见。

京东商城界面如图 3-3 所示，可以看到其中有很多个产品区块是完全相同的。假如纯粹使用 HTML 来写，然后遍历，则其他页面再有相同的区块时这些代码还得再重复一次，因此，组件化开发成了 Vue.js 项目中不可或缺的一个解决方案。

组件化开发中的组件分为全局组件与局部组件。全局组件通常可以在多个实例中共享，而局部组件只能在当前组件中调用。

### 3.2.2 全局组件

全局组件需要使用 Vue.component() 来定义，以下封装一个最简单的按钮组件，以此阐述全局组件的封装过程，代码如下：

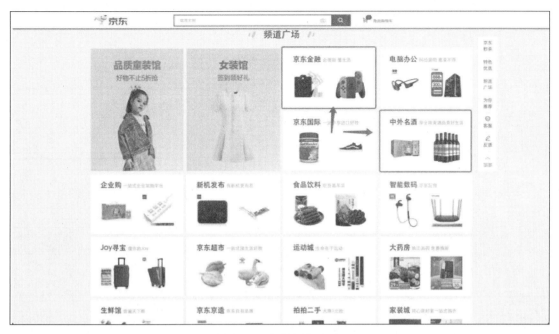

图3-3 京东商城界面

```
<!-- 第3章过滤器及组件化开发：Vue.component 实现按钮组件 -->
<!DOCTYPE html>
<html>
<head>
    <meta charset="UTF-8">
    <meta name="viewport" content="width=device-width, initial-scale=1.0">
    <meta http-equiv="X-UA-Compatible" content="ie=edge">
    <title></title>
</head>
<body>
<div id="app">
  <!-- 可以双标签调用 -->
    <my-btn></my-btn>
  <!-- 也可以单标签调用 -->
    <my-btn />
</div>
</body>
</html>
<script src="./Vue.js"></script>
<script>
Vue.component('my-btn', {
  template: `<button>这是一个按钮</button>`
})
new Vue({
    el: "#app",
```

            })
        </script>

至此，一个自定义的全局组件就诞生了。这里需要了解一点，template 中的代码就是 HTML 标签，开发者也可以将它抽离出来写，代码如下：

```
<!-- 第 3 章过滤器及组件化开发：自定义全局组件 template 的抽离 -->
<!DOCTYPE html>
<html>
<head>
    <meta charset="UTF-8">
    <meta name="viewport" content="width=device-width, initial-scale=1.0">
    <meta http-equiv="X-UA-Compatible" content="ie=edge">
    <title></title>
</head>
<body>
<div id="app">
    <my-btn />
</div>
</body>
</html>
<!-- 定义 template 模板 -->
<template id="temp">
   <button>这是一个按钮</button>
</template>
<script src="./Vue.js"></script>
<script>
Vue.component('my-btn', {
  template: "#temp"//组件使用模板
})
new Vue({
    el: "#app",
})
</script>
```

**注意** 实例中的所有属性与方法组件中都有。实例与组件只有两处不同：一是实例中的 el 在组件中需要使用 template 替代；二是实例中 data 是一个属性，而在组件中，data 必须是函数，因为每个组件需要复制并维护一份独立的对象数据，从而避免组件之间数据耦合或冲突。

### 1. 组件中的 data

在 Vue.js 组件中，data 必须是函数。函数需要返回一个对象，并在对象中存放必要的数据。实现图 3-4 所示的效果，同时要求按钮中的文字必须放在组件的 data 中，代码如下：

```
<!-- 第 3 章过滤器及组件化开发：Vue 组件中 data 必须是函数 -->
<!DOCTYPE html>
<html>
```

```html
<head>
    <meta charset="UTF-8">
    <meta name="viewport" content="width=device-width, initial-scale=1.0">
    <meta http-equiv="X-UA-Compatible" content="ie=edge">
    <title></title>
</head>
<body>
<div id="app">
    <my-btn />
</div>
</body>
</html>
<!-- 定义template模板 -->
<template id="temp">
  <button>{{msg}}</button>
</template>
<script src="./Vue.js"></script>
<script>
Vue.component('my-btn', {
  template: "#temp",//组件使用模板
  data(){//在组件中定义data
     return {
         msg: "This is a button."
     }
  }
})
new Vue({
    el: "#app",
})
</script>
```

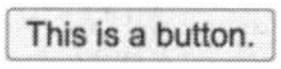

This is a button.

图 3-4　按钮中的文字动态存放于 data 中

### 2. 组件中的方法

对以上代码进行修改，给按钮添加一种方法，单击之后可以修改 msg 的值，从而让按钮的文字显示为"这是一个按钮"，代码如下：

```html
<!-- 第 3 章过滤器及组件化开发：组件中的方法 -->
<!DOCTYPE html>
<html>
<head>
    <meta charset="UTF-8">
    <meta name="viewport" content="width=device-width, initial-scale=1.0">
    <meta http-equiv="X-UA-Compatible" content="ie=edge">
    <title></title>
```

```html
</head>
<body>
<div id="app">
    <my-btn />
</div>
</body>
</html>
<!-- 定义template模板 -->
<template id="temp">
  <button @click="btnClick">{{msg}}</button>
</template>
<script src="./Vue.js"></script>
<script>
Vue.component('my-btn', {
  template: "#temp",
  data(){//定义组件数据
      return {
          msg: "This is a button."
      }
  },
  methods: {//定义组件方法
      btnClick(){
          this.msg = "这是一个按钮";
      }
  }
})
new Vue({
    el: "#app",
})
</script>
```

组件中除了 template 和 data 的写法与实例的写法不一样，其他写法都是完全一致的，例如 methods、computed、filters 等，因此初学者可以更快更容易地上手组件。

**3. 组件无法直接调用实例中的数据**

一个组件是无法直接调用实例中的数据的，除非实例愿意共享这部分数据。组件默认只能调用自身 data 中返回的数据，如果直接调用实例中的数据，则会出现如图 3-5 所示的错误提示信息。具体错误代码如下：

```html
<!-- 第3章过滤器及组件化开发：组件无法直接调用实例中的数据 -->
<!DOCTYPE html>
<html>
<head>
    <meta charset="UTF-8">
    <meta name="viewport" content="width=device-width, initial-scale=1.0">
    <meta http-equiv="X-UA-Compatible" content="ie=edge">
    <title></title>
```

```
</head>
<body>
<div id="app">
    <my-btn />
</div>
</body>
</html>
<!-- 定义 template 模板 -->
<template id="temp">
  <button>{{message}}</button>
</template>
<script src="./Vue.js"></script>
<script>
Vue.component('my-btn', {//添加全局组件
   template: "#temp"
})
new Vue({
    el: "#app",
    data: {
       message: "这是实例中的文字"
    }
})
</script>
```

图 3-5　组件无法直接调用实例中的数据

## 3.2.3　局部组件

全局组件可以在多个实例中共享，而局部组件只能在当前实例（或组件）中使用。

**注意**　组件除了可以在实例中调用，还可以在另一个组件中调用，也就是说一个组件可以调用另一个组件。

在组件或实例中，可以注册局部组件，代码如下：

```
<!-- 第 3 章过滤器及组件化开发：注册局部组件 -->
<!DOCTYPE html>
<html>
<head>
```

```html
    <meta charset="UTF-8">
    <meta name="viewport" content="width=device-width, initial-scale=1.0">
    <meta http-equiv="X-UA-Compatible" content="ie=edge">
    <title></title>
</head>
<body>
<div id="app">
    <my-btn />
</div>
</body>
</html>
<template id="temp">
  <button>{{msg}}</button>
</template>
<script src="./Vue.js"></script>
<script>
new Vue({
    el: "#app",
    components: {
        "my-btn": {//添加局部组件
            template: "#temp",
            data(){
                return {
                    msg: "局部组件按钮"
                }
            }
        }
    }
})
</script>
```

使用 components 属性可以定义（或者称为"注册"）一个组件，其格式与全局组件的格式大致相同，依然拥有 template、data 等属性与方法。

### 1. 组件通信之"父传子"

实例与组件、组件与组件之间是可以相同通信的，如果一个组件是被调用的，则习惯上将其称为子组件，而调用它的组件（或实例），就被称为父组件。父组件的数据子组件无法直接调用，必须通过"父传子"才能实现，代码如下：

```html
<!-- 第3章过滤器及组件化开发：组件通信之"父传子" -->
<!DOCTYPE html>
<html>
<head>
    <meta charset="UTF-8">
    <meta name="viewport" content="width=device-width, initial-scale=1.0">
    <meta http-equiv="X-UA-Compatible" content="ie=edge">
    <title></title>
```

```
</head>
<body>
<div id="app">
  <!-- 在标签上通过 v-bind 绑定 msg1,并赋值 msg,传递给子组件 -->
    <my-btn v-bind:msg1="msg" />
</div>
</body>
</html>
<!-- 定义组件模板 -->
<template id="temp">
  <button>{{msg1}}</button>
</template>
<script src="./Vue.js"></script>
<script>
new Vue({
    el: "#app",
    data: {
        msg: "父组件的信息"
    },
    components: {//在 components 下添加自定义组件
        "my-btn": {
            template: "#temp",
            //使用 props 属性,以数组形式接收父组件传递过来的数据
            props: ['msg1']
        }
    }
})
</script>
```

**注意** 此处有个细节需要留意,传递的属性作为子组件的值来使用,此时子组件 data 中不能重复定义相同名称的属性。

#### 2. props 的两种格式

props 有两种格式,一种是简写的数组格式,另一种是完整的对象格式。若要修改为对象格式,则代码如下:

```
//props 的两种格式
components: {
   "my-btn": {
      template: "#temp",
      //使用 props 属性,以对象的形式接收父组件传递过来的数据
      props: {
         msg1: {
             //使用 type 来确定传入的值的类型
             type: String,
             //使用 default 来定义默认的数据
```

```
            default: ""
        }
    }
}
```

当父组件并未给子组件传值时，建议添加上默认值，避免出现渲染缺失等错误。

### 3. 单项数据流

正如 Vue.js 官网所书："所有的 prop 都使其父子 prop 之间形成了一个单向下行绑定：父级 prop 的更新会向下流动到子组件中，但是反过来则不行。这样会防止从子组件意外变更父级组件的状态，从而导致应用的数据流向难以理解。"

因此可以得出结论：Vue.js 的父传子为单向传递，为了避免数据错乱，开发中应减少甚至不允许子组件直接修改 props 接收的数据，但有时在实际开发需求中，确实需要在子组件修改父组件传递过来的数据，此种情况应该怎么办呢？

### 4. 组件通信之"子传父"

笔者以最简单的方式对上文中抛出的问题给出答案，即子组件修改父组件数据的唯一途径是触发父组件去修改自身的数据，代码如下：

```html
<!-- 第 3 章过滤器及组件化开发：组件通信之"子传父" -->
<!DOCTYPE html>
<html>
<head>
    <meta charset="UTF-8">
    <meta name="viewport" content="width=device-width, initial-scale=1.0">
    <meta http-equiv="X-UA-Compatible" content="ie=edge">
    <title></title>
</head>
<body>
<div id="app">
    <!-- 使用@+方法名来接收子传父的函数 -->
    <my-btn :msg1="msg" @changemsg="changemsg1" />
</div>
</body>
</html>
<!-- 定义组件模板 -->
<template id="temp">
    <div>
        <h3>{{msg1}}</h3>
        <button @click="changeMsg">单击修改 msg1</button>
    </div>
</template>
<script src="./Vue.js"></script>
<script>
new Vue({
```

```
        el: "#app",
        data: {
            msg: "你好世界"
        },
        components: {
            "my-btn": {<!-- 注册组件 -->
                template: "#temp",
                props: ['msg1'],
                methods: {
                    changeMsg(){
                        //使用$emit向父组件发送一个函数名称,以及一个参数
                        this.$emit("changemsg", "hello world")
                    }
                }
            }
        },
        methods: {
            //被子组件触发的方法
            changemsg1(arg){
                this.msg = arg;
            }
        }
    })
</script>
```

#### 5. 同级组件通信

同级组件相互通信,目前有两种方式,一种是中央事件总线,另一种是状态管理 Vuex。由于 Vuex 的用法相对复杂,后文再行阐述,本章先介绍中央事件总线。

中央事件总线被称为 EventBus（也称为事件巴士）,代码如下：

```
<!-- 第3章过滤器及组件化开发：同级组件通信之中央事件总线 -->
<!DOCTYPE html>
<html>
<head>
    <meta charset="UTF-8">
    <meta name="viewport" content="width=device-width, initial-scale=1.0">
    <meta http-equiv="X-UA-Compatible" content="ie=edge">
    <title></title>
</head>
<body>
<div id="app">
    <my-btn1></my-btn1>
    <my-btn2></my-btn2>
</div>
</body>
</html>
<!--定义组件模板1-->
```

```html
<template id="temp1">
    <button @click="btnClick">按钮1</button>
</template>
<!--定义组件模板2-->
<template id="temp2">
    <button>按钮2</button>
</template>
<script src="./Vue.js"></script>
<script>
let bus = new Vue();

new Vue({
    el: "#app",
    data: {
        msg: "你好世界"
    },
    components: {
        "my-btn1": {//注册组件1
            template: "#temp1",
            methods: {
                btnClick(){
                    //使用bus.$emit发送一个函数名称，以及一个参数
                    bus.$emit("btn", "Hello World");
                }
            }
        },
        "my-btn2": {//注册组件2
            template: "#temp2",
            //在mounted中使用bus.$on接收
            mounted(){
                bus.$on("btn", (val)=>{
                    console.log(`按钮1传递过来的数据为${val}`);
                })
            }
        }
    }
})
</script>
```

单击了按钮1，按钮2的组件会得到按钮1被单击后传递过来的值，效果如图3-6所示。

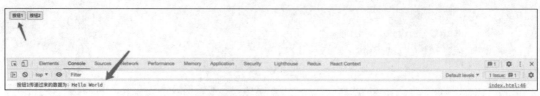

图3-6　EventBus实现同级组件通信

Vue.js 的核心思想之一便是本章所讲解的组件化开发，其中包括全局组件、局部组件、组件通信与组件调用等重要内容。

## 3.3　书店购物车项目实战

本节通过一个小型项目（书店购物车）来帮助读者掌握前文所提到的 Vue.js 相关知识点，效果如图 3-7 所示。

| 编号 | 书籍名称 | 出版日期 | 价格 | 购买数量 | 操作 |
|---|---|---|---|---|---|
| 1 | 算法导论 | 2006-9 | ¥ 85.00 元 | - 1 + | 移除 |
| 2 | UNIX编程艺术 | 2006-2 | ¥ 59.00 元 | - 1 + | 移除 |
| 3 | Vue程序设计 | 2008-10 | ¥ 35.00 元 | - 1 + | 移除 |
| 4 | 颈椎康复 | 2006-3 | ¥ 129.00 元 | - 1 + | 移除 |

总价格：¥ 308.00 元

图 3-7　书店购物车案例

### 1. 搭建 table 结构

首先，搭建 table 结构，后续真实的数据需要靠 for 循环渲染。table 结构使用 HTML+CSS 实现，代码如下：

```
<!-- 第 3 章过滤器及组件化开发：书店购物车 table 结构搭建 -->
<!DOCTYPE html>
<html>
<head>
    <meta charset="UTF-8">
    <meta name="viewport" content="width=device-width, initial-scale=1.0">
    <meta http-equiv="X-UA-Compatible" content="ie=edge">
    <title>购物车表格</title>
    <style>
    table{width: 800px;margin: 100px auto;text-align: center;border-collapse:
collapse;}
    th,td{border: 1px solid #000;padding: 4px 0;}
    tfoot{text-align-last: left;}
    tfoot td{border: 0;}
    </style>
</head>
<body>
<div id="app">
    <!-- 表格结构 -->
    <table>
      <thead>
          <tr>
              <th>编号</th>
              <th>书籍名称</th>
```

```html
                <th>出版日期</th>
                <th>价格</th>
                <th>购买数量</th>
                <th>操作</th>
            </tr>
        </thead>
        <tbody>
            <tr>
                <td>1</td>
                <td>算法导论</td>
                <td>2006-9</td>
                <td>¥ 85.00</td>
                <td>
                    <button>-</button>
                    <span>1</span>
                    <button>+</button>
                </td>
                <td><button>移除</button></td>
            </tr>
        </tbody>
        <tfoot>
            <tr>
                <td colspan="6">总价格：</td>
            </tr>
        </tfoot>
    </table>
</div>
</body>
</html>
<script src="./Vue.js"></script>
<script>
new Vue({
    el: "#app",
})
</script>
```

运行结果如图 3-8 所示。

| 编号 | 书籍名称 | 出版日期 | 价格 | 购买数量 | 操作 |
|---|---|---|---|---|---|
| 1 | 算法导论 | 2006-9 | ¥ 85.00 | - 1 + | 移除 |
| 总价格： | | | | | |

图 3-8  table 结构搭建

### 2. 数据渲染

本项目需要数据渲染，此处提供循环的数据。分别有表格表头和表体数据，代码如下：

```
//用于渲染的表格数据
```

```
data: {
//表头数据
titles: ['编号', '书籍名称', '出版日期', '价格', '购买数量', '操作'],
//表体数据
books: [
    {
      id: 1,
      name: '算法导论',
      date: '2006-9',
      price: 85,
      num: 1
    },
    {
      id: 2,
      name: 'UNIX 编程艺术',
      date: '2006-2',
      price: 59,
      num: 1
    },
    {
      id: 3,
      name: 'Vue 程序设计',
      date: '2008-10',
      price: 35,
      num: 1
    },
    {
      id: 4,
      name: '颈椎康复',
      date: '2006-3',
      price: 129,
      num: 1
    }
  ]
}
```

接下来实现 for 循环遍历，首先渲染表格头部，代码如下：

```
<!-- for 循环渲染表头 -->
<thead>
    <tr>
        <th v-for="(item,index) in titles" :key="index">{{item}}</th>
    </tr>
</thead>
```

表头渲染的结果如图 3-9 所示。

| 编号 | 书籍名称 | 出版日期 | 价格 | 购买数量 | 操作 |
|---|---|---|---|---|---|

图 3-9  表头渲染

表头渲染结束后，接下来实现表体渲染，表体的渲染是针对 tr 的渲染，代码如下：

```html
<!-- for 循环渲染表体 -->
<tbody>
    <tr v-for="(item,index) in books" :key="item.id">
        <td>{{item.id}}</td>
        <td>{{item.name}}</td>
        <td>{{item.date}}</td>
        <td>{{item.price}}</td>
        <td>
            <button>-</button>
            <span>{{item.num}}</span>
            <button>+</button>
        </td>
        <td><button>移除</button></td>
    </tr>
</tbody>
```

运行结果如图 3-10 所示。

| 编号 | 书籍名称 | 出版日期 | 价格 | 购买数量 | 操作 |
|---|---|---|---|---|---|
| 1 | 算法导论 | 2006-9 | 85 | - 1 + | 移除 |
| 2 | UNIX编程艺术 | 2006-2 | 59 | - 1 + | 移除 |
| 3 | Vue程序设计 | 2008-10 | 35 | - 1 + | 移除 |
| 4 | 颈椎康复 | 2006-3 | 129 | - 1 + | 移除 |

总价格：

图 3-10  表体渲染

### 3. 价格过滤器

本项目的价格都是一个纯数字，如果希望实现如图 3-7 所示的价格格式，则需要使用过滤器。全局过滤器和局部过滤器均可，此处选用局部过滤器，代码如下：

```javascript
//价格过滤器
new Vue({
    el: "#app",
    data: {...},
    //此处省略 data 的代码
    filters: {
        priceFilter(val){
            return "¥"+Number(val).toFixed(2)+" 元";
        }
    }
})
```

构建完过滤器后，在标签中使用，代码如下：

```
<td>{{item.price | priceFilter}}</td>
```

运行结果如图 3-11 所示。

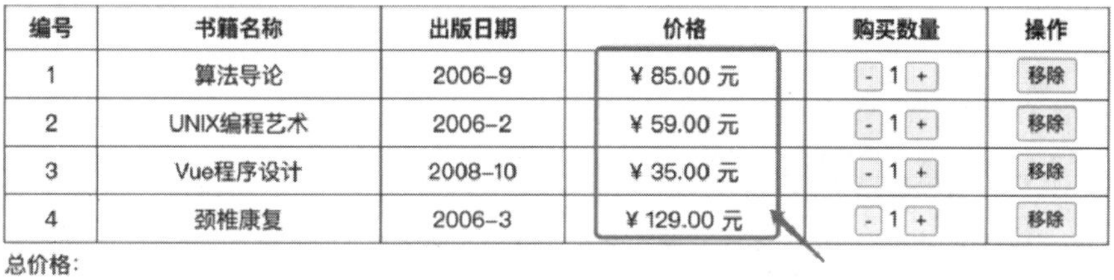

图 3-11　过滤器

### 4. 计算总价

总价的计算需要遍历整个 books 数组，将每项的价格与数量相乘并累加。这里选用 JavaScript 的 reduce()方法，代码如下：

```
computed: {
    totalPrice(){
        return this.books.reduce((total, item)=>{
            return total += item.price * item.num;
        }, 0)
    }
}
```

在 HTML 中调用并启用价格过滤器，代码如下：

```
<tfoot>
  <tr>
    <td colspan="6">总价格：{{totalPrice | priceFilter}}</td>
  </tr>
</tfoot>
```

运行结果如图 3-12 所示。

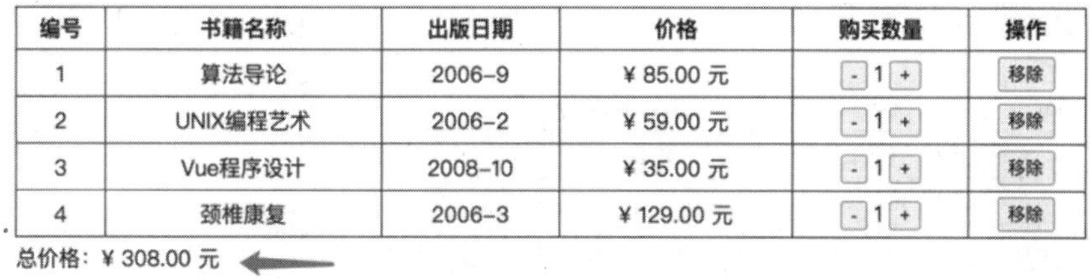

图 3-12　计算总价

### 5. 购买数量

购买数量可以增加或减少。这里主要的逻辑在于：当触发减少时，如果数量小于或等于

1,就需要禁止减少,因此需要增加判断语句。同时,商品数量的增加和减少都会使总价自动发生变化,购买数量的代码如下:

```html
<!-- 处理购买数量的事件 -->
<tbody>
    <tr v-for="(item,index) in books" :key="item.id">
        <td>{{item.id}}</td>
        <td>{{item.name}}</td>
        <td>{{item.date}}</td>
        <td>{{item.price | priceFilter}}</td>
        <td>
            <button @click="decrease(index)">-</button>
            <span>{{item.num}}</span>
            <button @click="item.num++">+</button>
        </td>
        <td><button>移除</button></td>
    </tr>
</tbody>

<!-- 以下为JS代码 -->
<script>
new Vue({
    el: "#app",
    methods: {
        //减少
        decrease(index){
            this.books[index].num--;
        }
    }
})
</script>
```

运行结果如图3-13所示。

| 编号 | 书籍名称 | 出版日期 | 价格 | 购买数量 | 操作 |
|---|---|---|---|---|---|
| 1 | 算法导论 | 2006-9 | ¥ 85.00 元 | - 2 + | 移除 |
| 2 | UNIX编程艺术 | 2006-2 | ¥ 59.00 元 | - 1 + | 移除 |
| 3 | Vue程序设计 | 2008-10 | ¥ 35.00 元 | - 2 + | 移除 |
| 4 | 颈椎康复 | 2006-3 | ¥ 129.00 元 | - 1 + | 移除 |

总价格:¥ 428.00 元

图3-13 购买数量的增加与减少

### 6. 移除商品

移除商品就是删除指定索引值的数组项,代码如下:

```html
<!-- 以下为HTML代码 -->
```

```
<td><button @click="remove(index)">移除</button></td>

<!-- 以下为JavaScript代码 -->
<script>
new Vue({
    methods: {
        //移除索引值对应的数组项
        remove(index){
            this.books.splice(index, 1);
        }
    }
})
</script>
```

至此，书店购物车项目完成。

本项目通过购物车操作示例，帮助读者回顾循环渲染、事件、过滤器、计算属性等 Vue 核心知识点。

# 第 4 章 Webpack、Slot 与 Vue CLI 脚手架

## 4.1 Webpack 模块化打包工具

随着前端的发展,开发者往往会对一个项目进行工程化、模块化及自动化,而目前 Webpack 就是这一领域的代表。本章主要讲述 Webpack,与 Vue.js 暂时没有关系,但自本章之后的所有章节,Vue.js 的所有知识点将基于 Webpack。

### 4.1.1 Webpack 的简介与安装

**1. Webpack 简介**

从本质上来讲,Webpack 是一个静态模块打包工具。要想让项目中写好的模块化代码在各式各样的浏览器上能够兼容,就必须借助于其他工具,而 Webpack 的其中一个核心就是让开发人员可以进行模块化开发,并处理模块间的依赖关系。不仅是 JavaScript 文件、CSS、图片、JSON 文件等在 Webpack 中都可以当作模块来使用,这就是 Webpack 的模块化概念。

**2. Webpack 安装**

首先新建一个空白文件夹(本章将其命名为 demo)作为项目,在项目的根目录下打开 cmd,并执行项目初始化,代码如下:

```
#执行后生成package.json 文件
npm init -y
```

注意 本章后续所有的命令行操作均在根目录下操作,后文将不再强调。

1)依赖安装

初始化项目后,可以进行 Webpack 安装,本章还需要借助 jQuery 框架实现部分示例,因此可以把 jQuery 一起安装,代码如下:

```
npm install webpack webpack-cli -D
npm install jquery -S
```

**注意** 在上述命令行中，-D 表示安装到 package.json 的开发依赖 devDependencies（开发环境）对象里，也可以用--save-dev 代替，而-S 是--save 的简写，这样安装会安装到 dependencies（生产环境）对象里，也可以用--save 代替。

安装完成后，可以看到初始化产生的 package.json 文件，代码如下：

```
//第 4 章 Webpack、Slot 与 Vue CLI 脚手架：package.json 文件
{
  "name": "demo",
  "version": "1.0.0",
  "description": "",
  "main": "index.js",
  "scripts": {
    "test": "echo \"Error: no test specified\" && exit 1"
  },
  "author": "",
  "license": "ISC",
  "devDependencies": {
    "webpack": "^5.58.1",
    "webpack-cli": "^4.9.0"
  },
  "dependencies": {
    "jquery": "^3.6.0"
  }
}
```

**注意** devDependencies 与 dependencies 的区别在于：在发布 npm 包时，本身 dependencies 下的模块会作为依赖，一起被下载；devDependencies 下面的模块就不会自动下载了，但对于项目而言，npm install 会自动下载 devDependencies 和 dependencies 下面的模块。

**2）创建入口文件**

在项目的根目录下创建 index.html 文件，并写一个列表结构，代码如下：

```html
<!--第 4 章 Webpack、Slot 与 Vue CLI 脚手架：入口文件 -->
<!DOCTYPE html>
<html lang="en">
<head>
  <meta charset="UTF-8">
  <meta name="viewport" content="width=device-width, initial-scale=1.0">
  <meta http-equiv="X-UA-Compatible" content="ie=edge">
  <title>Document</title>
</head>
<body>
  <ul>
    <li></li>
    <li></li>
```

```
        <li></li>
        <li></li>
        <li></li>
        <li></li>
        <li></li>
    </ul>
</body>
</html>
```

然后在项目的根目录下创建 src 目录，再到 src 目录中创建 index.js 文件，在其中写入 jQuery 代码，用于实现列表隔行换色，代码如下：

```
import $ from 'jquery'
$('ul li:even').css({background: 'red'})
$('ul li:odd').css({background: 'green'})
```

到此，理论上已经可以实现隔行换色，但当浏览器打开后，真实的情况是出现了报错，如图 4-1 所示。

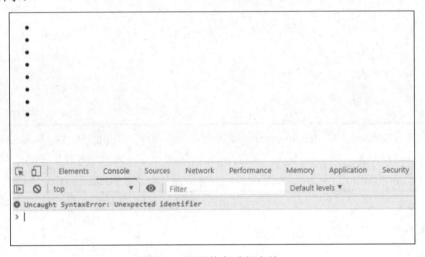

图 4-1　隔行换色功能失效

为什么会报错？因为浏览器并不兼容以 import 引入模块这种方式，所以这里需要用到 Webpack 进行打包。

### 4.1.2　Webpack 基本配置

在项目根目录下，新建 webpack.config.js 文件，这是 Webpack 的项目配置文件，此处可以在其中对入口文件、出口文件等配置进行定义，代码如下：

```
//第 4 章 Webpack、Slot 与 Vue CLI 脚手架：Webpack 配置文件
const path = require('path');
```

```
module.exports = {
  mode: "development",
  //dirname 代表索引到文件所在目录
  entry: path.join(__dirname, './src/index.js'),
  output: {
    path: path.join(__dirname, './dist'),
    filename: 'main.js'
  }
}
```

然后在 package.json 文件中配置打包命令,代码如下:

```
"scripts": {
    "build": "webpack --config webpack.config.js"
}
```

修改 index.html 文件,代码如下:

```
<script src="./dist/main.js"></script>
```

再用浏览器打开 index.html 文件查看效果,即可看到如图 4-2 所示的隔行换色效果。

图 4-2 隔行换色效果

### 4.1.3 webpack-dev-server

虽然目前实现了隔行换色,但这时如果将 index.html 文件实现的背景颜色由 green 改成 pink,则会发现即使浏览器刷新也没有效果,需要再运行一次 npm run start 命令才有用,这时就需要 webpack-dev-server(热重载)。

想要使用 webpack-dev-server,需要先进行安装,打开命令行,执行如下代码:

```
npm install webpack-dev-server -D
```

然后修改 package.json 文件中的项目启动命令,代码如下:

```
"scripts": {
  "start": "webpack-dev-server --open --port 3002 --hot",
  "build": "webpack --config webpack.config.js"
}
```

在以上代码中,--open 表示项目运行后自动打开浏览器,--port 表示服务监听端口,此处监听 3002 端口，--hot 表示自动热更新。

> **注意** 启动 webpack-dev-server 后，开发者在目标文件夹中是看不到编译后的文件的，实时编译后的文件都保存到了内存当中。如果开发者想看到 main.js 文件，则可以运行 http://localhost:3002/main.js 进行查看。

接下来，在 index.html 文件中可以对打包后的 main.js 文件进行引入，应当修改 JS 的路径，代码如下：

```html
<script src="./main.js"></script>
```

至此，通过 npm run start 命令开发者仍然无法访问 HTML 页面，因为打包出来的 dist 被隐藏，并且也不包含 index.html 文件，所以会有 Cannot GET / 的警告，因此，还需要解决 HTML 的问题。

### 4.1.4 html-webpack-plugin

首先安装 html-webpack-plugin，命令如下：

```
npm install html-webpack-plugin -D
```

然后修改 webpack.config.js 文件，代码如下：

```javascript
//第 4 章 Webpack、Slot 与 Vue CLI 脚手架：html-webpack-plugin 的使用
const path = require('path');
const HtmlWebpackPlugin = require('html-webpack-plugin');

module.exports = {
  //development 表示开发环境，production 表示生产环境
  mode: "development",
  entry: path.join(__dirname, './index.js'),
  output: {
    path: path.join(__dirname, './dist'),
    filename: 'main.js'
  },
  plugins: [
    new HtmlWebpackPlugin({
      template: path.join(__dirname, './index.html')
    })
  ]
}
```

接下来可以删掉 index.html 文件里面的 main.js 引用，因为 html-webpack-plugin 会自动把打包出来的 bundle 加到 index.html 代码里。重新运行 npm run start 命令，即可看到如图 4-3 的效果。

图 4-3　隔行换色热更新

除此之外，还可以随意更改 jQuery 代码中的颜色代码，页面也会随之热更新。

### 4.1.5　loader

loader 是 Webpack 用来对模块进行预处理的，在一个模块被引入之前，会预先使用 loader 处理模块的内容。

#### 1. css-loader 与 style-loader

传统的 CSS 写法是在 HTML 文件中使用 link 标签引入 CSS 代码，借助 Webpack 的 style-loader 和 css-loader 可以在入口 JS 文件中引入 CSS 文件并让样式生效。css-loader 是用来加载以.css 结尾的文件的，style-loader 的作用是使用<style>将 css-loader 内部样式注入 HTML 页面。

首先需要安装 css-loader 与 style-loader，命令如下：

```
npm install css-loader style-loader -D
```

在 webpack.config.js 文件中进行 loader 配置，代码如下：

```
//第4章 Webpack、Slot与Vue CLI脚手架：loader 配置
const path = require('path');
const HtmlWebpackPlugin = require('html-webpack-plugin');

module.exports = {
  //development 表示开发环境，production 表示生产环境
  mode: "development",
  entry: path.join(__dirname, './index.js'),
  output: {
    path: path.join(__dirname, './dist'),
    filename: 'bundle.js'
  },
  plugins: [
    new HtmlWebpackPlugin({
      template: path.join(__dirname, './index.html'),
      filename: 'index.html'
    })
  ],
  module: {
    rules: [{
```

```
    test: /\.css$/,
    //注意：这里的数组是反向读取的(从右往左)
    use: ['style-loader', 'css-loader']
  }]
}
```

在 src 目录下创建 index.css 文件，书写一些样式类代码，代码如下：

```
body {
  background: skyblue;
}
```

然后在 index.js 文件中引入 index.css 文件，并且重新执行 npm run start 命令，即可在修改 CSS 时看到页面样式的更新。

#### 2. less-loader

如果开发者的项目使用了类似于 Less 这种预编译语言，则还需要安装并配置 less-loader，命令如下：

```
npm install less-loader less -D
```

修改 webpack.config.js 文件中的 module，代码如下：

```
//第 4 章 Webpack、Slot 与 Vue CLI 脚手架：module 调用
module: {
  rules: [{
    test: /\.css$/,
    //注意：这里的数组是反向读取的(从右往左)
    use: ['style-loader', 'css-loader']
  },{
    test: /.less$/,
    use: ['style-loader', 'css-loader', 'less-loader']
  }]
}
```

### 4.1.6　babel

在真实项目开发中，前端开发人员往往不止使用 ES5，而是使用 ES6+ES5 同时进行开发，而 ES6 在低版本浏览器中存在兼容问题，因此需要使用 babel 将其转换为 ES5。

babel 是非常有名的 Webpack 插件，需要先安装（安装的插件比较多），命令如下：

```
npm install babel-loader @babel/core @babel/plugin-proposal-class-properties @babel/plugin-transform-runtime @babel/preset-env @babel/runtime -D
```

然后需要对 webpack.config.js 进行配置，代码如下：

```
//第 4 章 Webpack、Slot 与 Vue CLI 脚手架：babel 配置
```

```
module: {
  rules: [{
    test: /\.js$/,
    use: [{
      loader: 'babel-loader',
      options: {
        presets: [
          '@babel/preset-env'
        ],
        plugins: [
          [require("@babel/plugin-transform-runtime"), { "legacy": true }],
          [require("@babel/plugin-proposal-class-properties"),{"legacy": true}]
        ]
      }
    }],
    Excelude: /node_modules/
  }]
}
```

注意　Excelude 表示排除掉 node_modules 下载的依赖项。

接下来测试 babel，在 src 目录下的 index.js 文件中引入 ES6 语法，代码如下：

```
const fn = () => {
  console.log(123);
}

fn();
```

重新在命令行执行 npm run start 命令运行项目，打开内存中的 bundle.js 文件，可以看到这段代码的编译结果，最终代码如下：

```
var fn = function fn() {\n    console.log(123);\n}
```

## 4.1.7　HTML 热更新

目前本项目的 Webpack 配置只实现 JS 和 CSS 的热更新，假如开发者想要实现 HTML 的热更新，就需要使用 raw-loader，命令如下：

```
npm install --save-dev raw-loader
```

在 index.js 文件中引入 index.html 文件，代码如下：

```
import '../index.html';
```

然后在 webpack.config.js 文件中配置 raw-loader，代码如下：

```
//第 4 章 Webpack、Slot 与 Vue CLI 脚手架：raw-loader 配置
module.exports = {
```

```
    //此处省略上文中提到的一些配置
    ...
    module: {
      rules: [
        {
          test: /\.(htm|html)$/,
          use: [
            'raw-loader'
          ]
        },
        //此处省略上文中提到的一些配置
        ...
      ]
    }
}
```

重新执行 npm run start 命令即可实现 HTML 的热更新。

### 4.1.8 图片资源

Webpack 搭建的项目中也需要加载图片资源，可以在 webpack.config.js 文件中进行配置，代码如下：

```
//第 4 章 Webpack、Slot 与 Vue CLI 脚手架：图片配置
module.exports = {
    //此处省略上文中提到的一些配置
    ...
    module: {
      rules: [{
        test: /\.(png|gif|svg|jpe?g)$/,
        /*
          asset/resource 表示把指定资源复制到对应的目录
          asset/inline   表示直接转 base64，不产生图片
        */
        //type: 'asset/inline',
        type: 'asset/resource',
        generator: {
          filename: 'img/[name].[hash:4][ext]'
        }
      },
      //此处省略上文中提到的一些配置
      ...
      ]
    }
}
```

然后修改 index.html 文件，代码如下：

```
<div class="logo"></div>
<img src="" alt="">
```

再在项目 src 目录下创建 images 目录，并到 index.css 文件中书写背景图片引入，代码如下：

```
.logo{
  width: ...;
  height: ...;
  background: url(./images/logo.png);
}
```

再到 index.js 文件中，使用 JS 操作图片引入，代码如下：

```
import logo from "./assets/images/logo.png"
document.getElementsByTagName('img')[0].src = logo;
```

运行 npm run start 命令即可看到图片，或者运行 npm run build 命令查看打包后的结果，并双击 dist/index.html 文件手动打开页面查看效果。至此，关于 Webpack 日常开发中高频使用的知识点便已讲述完毕。

本节主要讲解了 Webpack 的基本配置、loader 及 plugin 等主要内容。

## 4.2　Vue CLI

Vue CLI 俗称 Vue.js 脚手架，官网网址为 https://cli.vuejs.org/zh/ 。

### 4.2.1　Vue CLI 的简介与安装

**1. Vue CLI 简介**

Vue CLI 是一个基于 Vue.js 进行快速开发的完整系统，并致力于将 Vue.js 生态中的工具基础标准化。Vue CLI 确保了各种构建工具能够基于智能的默认配置即可平稳衔接，这样开发者可以专注在撰写应用上，而不必花时间去纠结配置问题。与此同时，它也为每个工具提供了调整配置的灵活性，而无须 eject（解包）。

**2. Vue CLI 安装**

首先，开发者应保证计算机的 Node.js 环境版本在 8.9 以上，当前建议安装 12.10.0 版本，该版本相对稳定，然后在计算机中的任意位置打开 cmd 进行全局安装，代码如下：

```
$ npm install -g @vue/cli
#OR
$ yarn global add @vue/cli
```

安装过后，开发者可以通过版本号检查，验证 Vue CLI 是否安装成功，代码如下：

```
$ vue --version
```

如果开发者的计算机曾经安装过 Vue CLI 的旧版本，则升级 Vue CLI 的代码如下：

```
$ npm update -g @vue/cli

#或者
$ yarn global upgrade --latest @vue/cli
```

### 4.2.2　Vue CLI 创建项目

使用 Vue CLI 创建项目的方式是通过命令行创建，代码如下：

```
#本章假设项目名称为 demo
$ vue create demo
```

创建后会出现一个配置选项界面，该界面用于选择项目中需要预安装的插件，如图 4-4 所示。

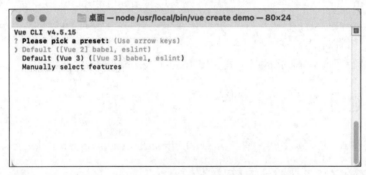

图 4-4　Vue CLI 选项界面

接下来将一步一步地带着 Vue CLI 初学者挑选配置项。一般开发者需要自定义选项，通过键盘的向上或向下键，选择第 3 个选项 Manually select features，然后进入"自选插件"界面，如图 4-5 所示。

图 4-5　"自选插件"界面

使用键盘向上或向下键可以移动到选项上，按下空格键可以选中该选项。笔者挑选了适合 Vue CLI 初学者的预安装插件，如图 4-5 所示。

选中以上选项后，按 Enter 键，进入"Vue.js 版本选择"界面，如图 4-6 所示。

图 4-6 "Vue.js 版本选择"界面

选择 2.x，按 Enter 键进入路由模式，按 Y 键，进入"预编译语言选项"界面，如图 4-7 所示。

图 4-7 "预编译语言选项"界面

选择 Less，按 Enter 键，进入"配置文件选项"界面，如图 4-8 所示。

选择 In package.json，按 Enter 键，进入"项目预设确认"界面，初学者可以根据自己的实际情况选择是否保留预设选项。此处笔者选择 Y，按 Enter 键，并输入预设选项的名称，按 Enter 键后命令行便开始创建项目。创建成功后，会进入提示创建成功的界面，如图 4-9 所示。

根据提示，在命令行输入 cd demo 命令，即可进入该项目的根目录，如果想要通过 Visual Studio Code 软件打开，则可以直接执行的命令如下：

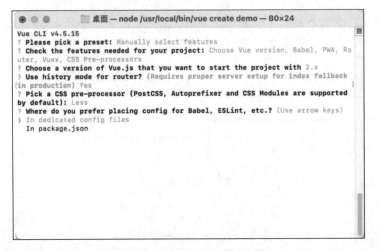

图 4-8 "配置文件选项"界面

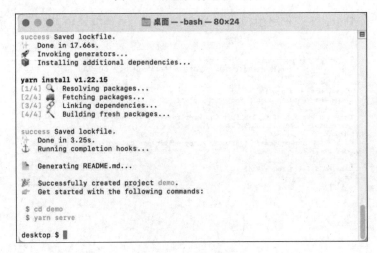

图 4-9 项目创建成功

```
$ cd demo
#用 VS Code 打开
$ code .
```

**注意** Visual Studio Code 是目前市面上前端开发软件中既免费，平台支持性又高的软件，是由微软公司研发的。

### 4.2.3 Vue CLI 项目预览

通过 VS Code 打开项目，如图 4-10 所示，底部的控制台可以使用快捷键 Ctrl+~打开。

第4章　Webpack、Slot与Vue CLI脚手架　59

图 4-10　VS Code 打开 Vue CLI 项目

想要运行该项目，只需在底部命令行执行如下代码：

```
$ yarn serve
```

运行效果如图 4-11 所示，表示服务器已经自动运行。

图 4-11　项目运行界面

按住 Ctrl 键与鼠标左键，可以在浏览器打开该项目进行预览。Vue CLI 运行起来的项目初始界面如图 4-12 所示。

图 4-12　项目初始界面预览

本节主要讲解了 Vue CLI 如何安装、创建并运行项目进行预览。

# 第 5 章 Vue.js 高级语法

## 5.1 插槽

对于 Slot（插槽）的概念，Vue.js 官网只简单地介绍了一句："Vue.js 实现了一套内容分发的 API，这套 API 的设计灵感源自 Web Components 规范草案，将<slot>元素作为承载分发内容的出口。"但这对于初学者而言，很难理解何为"内容分发"。

笔者以最直白的方式对 Slot 进行描述：当一个自定义组件以双标签形式调用时，写在标签中的任何元素与文本都需要借助 Slot 方能实现内容的显示。

### 5.1.1 匿名插槽

在子组件中，使用特殊的元素<slot>就可以为子组件开启一个插槽，该插槽插入什么内容取决于父组件如何使用。

首先，封装一个自定义组件 MyComp.vue，代码如下：

```
<!-- 第 5 章 Vue.js 高级语法：MyComp.vue -->
<template>
  <div>
     <h2>This is a custom component.</h2>
  </div>
</template>
```

引入到父组件后，将自定义的组件以双标签的形式书写，并且在自定义组件中写入内容，代码如下：

```
<!-- 第 5 章 Vue.js 高级语法：在双标签组件中写入内容 -->
<template>
  <div>
    <my-comp>
      <a href="https://www.baidu.com">跳转到百度</a>
    </my-comp>
  </div>
</template>
```

```
<script>
import MyComp from '@/components/MyComp'

export default {
  name: 'Home',
  components: {
    MyComp
  }
}
</script>
```

打开浏览器查看，会发现 a 标签并没有显示出来。想要实现 a 标签的显示，需要在 MyComp.vue 文件中写入<slot></slot>。slot 标签的位置决定了 a 标签的显示位置，代码如下：

```
<!-- MyComp.vue -->
<template>
  <div>
    <h2>This is a custom component.</h2>
    <slot></slot>
  </div>
</template>
```

最终 a 标签的显示效果如图 5-1 所示。

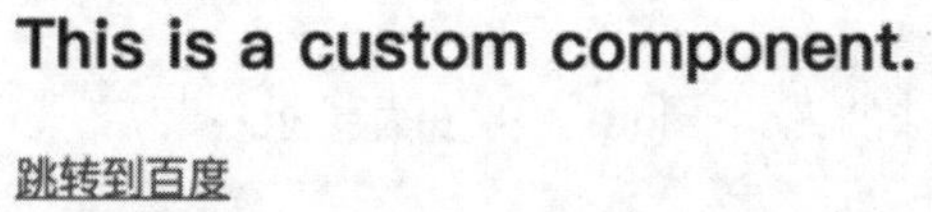

**This is a custom component.**

<u>跳转到百度</u>

图 5-1  a 标签显示于 h2 标签下方

## 5.1.2  具名插槽

上文中提到，在自定义组件中插入内容，可以使用匿名插槽将内容显示出来，但如果需求有所变更，在自定义组件中插入多个元素，则要求显示在不同位置。针对这个需求，又该如何处理呢？示例代码如下：

```
<!-- 在自定义组件中插入多个元素 -->
<template>
  <div>
    <my-comp>
      <span>要求显示在 h2 标签上方</span>
      <span>要求显示在 h2 标签下方</span>
    </my-comp>
  </div>
</template>
```

解决方案是使用具名插槽。给上述代码中的两个 span 添加属性 slot，指定 slot 名称，代码如下：

```
<template>
  <div>
    <my-comp>
      <span slot="up">要求显示在h2标签上方</span>
      <span slot="down">要求显示在h2标签下方</span>
    </my-comp>
  </div>
</template>
```

然后在自定义组件中为 slot 匹配名称,这样就可以决定插槽内容的显示位置,从而实现本章开头提及的"内容分发",代码如下:

```
<template>
  <div>
    <slot name="up"></slot>
    <h2>This is a custom component.</h2>
    <slot name="down"></slot>
  </div>
</template>
```

最终两个 span 的显示效果如图 5-2 所示。

要求显示在h2标签上方

**This is a custom component.**

要求显示在h2标签下方

图 5-2  具名插槽实现内容分发

### 5.1.3  作用域插槽

有时让插槽内容能够访问子组件中才有的数据是很有用的,例如想要在上述的两个 span 中使用 MyComp.vue 组件的数据。演示 MyComp.vue 组件的数据传递方式,代码如下:

```
<!-- 第 5 章 Vue.js 高级语法:将 MyComp.vue 文件中的数据传递给 slot 对应的元素 -->
<template>
  <div>
    <!--传入插槽数据-->
    <slot name="up" :upMsg="upMsg"></slot>
    <h2>This is a custom component.</h2>
    <slot name="down" :downMsg="downMsg"></slot>
  </div>
</template>

<script>
export default {
    data(){
```

```
        return {
            upMsg: "你好世界",//定义插槽数据
            downMsg: "Hello World"
        }
    }
}
</script>
```

而 slot 对应的元素调用数据的代码如下：

```
<my-comp>
  <template v-slot:up="scope">
    <!--接收并渲染插槽数据-->
    <span>{{scope.upMsg}}</span>
  </template>
  <template v-slot:down="scope">
    <span>{{scope.downMsg}}</span>
  </template>
</my-comp>
```

最终实现的效果如图 5-3 所示。

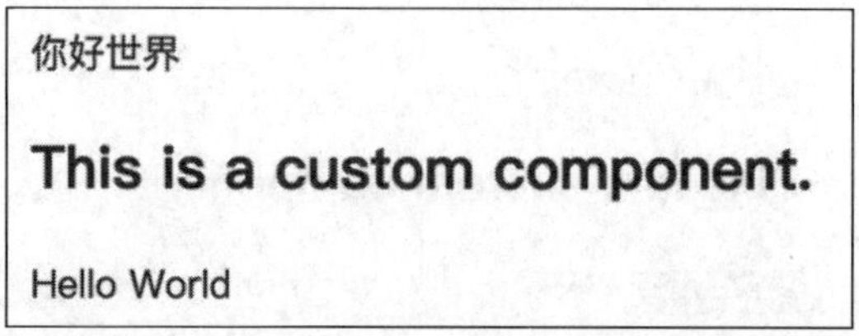

你好世界

This is a custom component.

Hello World

图 5-3　插槽对应的元素调用组件中的数据

---

注意　作用域插槽实际上有好几种写法，但其他写法（如 slot-scope）已在官方文档中写明"已废弃"，因此本书不对其他写法进行介绍。

---

本节主要讲解匿名插槽、具名插槽与作用域插槽的使用方法。

## 5.2　修饰符

Vue.js 提供了修饰符以帮助开发者简化代码，包括表单修饰符、事件修饰符及按键修饰符。

### 5.2.1　表单修饰符

一个表单的 v-model 指令可以携带修饰符。

#### 1. lazy

在默认情况下，v-model 在每次 input 事件触发后会对输入框的值与数据进行同步。开

发者可以添加 lazy 修饰符，从而转换为在 change()事件之后进行同步，代码如下：

```
<!-- 在"change"时而非"input"时更新 -->
<input v-model.lazy="msg" />
```

2. number

如果想自动将用户的输入值转换为数值类型，则可以给 v-model 添加 number 修饰符，代码如下：

```
<input v-model.number="age" type="number" />
```

这通常很有用，因为即使在 type="number" 时，HTML 输入元素的值也总会返回字符串。如果这个值无法被 parseFloat()解析，则会返回原始的值。

3. trim

如果要自动过滤用户输入的首尾空白字符，则可以给 v-model 添加 trim 修饰符，代码如下：

```
<input v-model.trim="msg" />
```

### 5.2.2 事件修饰符

在事件处理程序中调用 event.preventDefault()或 event.stopPropagation()是非常常见的需求。尽管开发者可以在方法中轻松实现，但更好的方式是：方法只有纯粹的数据逻辑，而不是去处理 DOM 事件细节。针对 v-on 所处理的事件，Vue.js 提供了一系列事件修饰符。

1. stop

stop 修饰符可以阻止单击事件继续传播，代码如下：

```
<a v-on:click.stop="doThis"></a>
```

2. prevent

prevent 修饰符可以提交事件并不再重载页面，代码如下：

```
<form v-on:submit.prevent="onSubmit"></form>
<!-- 修饰符可以串联 -->
<a v-on:click.stop.prevent="doThat"></a>
<!-- 也可以只有修饰符 -->
<form v-on:submit.prevent></form>
```

3. capture

capture 修饰符用于在添加事件监听器时使用事件捕获模式，即内部元素触发的事件先在此处理，然后交由内部元素进行处理，代码如下：

```
<div v-on:click.capture="doThis">...</div>
```

4. self

self 修饰符只当在 event.target 是当前元素自身时才触发处理函数，即事件不是从内部元素触发的，代码如下：

```
<div v-on:click.self="doThat">...</div>
```

#### 5. once

once 修饰符表示单击事件将只会触发一次，代码如下：

```
<a v-on:click.once="doThis"></a>
```

#### 6. passive

当一个滚动事件添加了 passive 修饰符后，滚动行为将会立即触发，不会等待 onScroll 完成，这其中包含 event.preventDefault()的情况，代码如下：

```
<div v-on:scroll.passive="onScroll">...</div>
```

**注意** 不要把 .passive 和 .prevent 一起使用，因为 .prevent 将会被忽略，同时浏览器可能会发出一个警告。

### 5.2.3 按键修饰符

在监听键盘事件时，经常需要检查详细的按键。Vue.js 允许为 v-on 在监听键盘事件时添加按键修饰符，代码如下：

```
<!-- 只有在 key 是 Enter 时调用 vm.submit() -->
<input v-on:keyup.enter="submit" />

<!-- 或者使用键码 -->
<input v-on:keyup.13="submit" />
```

本节主要讲解了 Vue.js 的修饰符，包括表单修饰符、事件修饰符和按键修饰符，这些修饰符将在开发过程中节省开发者的冗余代码，从而提升效率。

## 5.3 监听

开发者在项目中时常需要监听一个数据的变化，并根据这个变化去触发一些事件，此时可以使用 Vue.js 提供的 Watch 侦听器，俗称监听器。

### 5.3.1 普通监听

监听某个基础数据类型的变化，代码如下：

```
<!-- 第 5 章 Vue.js 高级语法：Vue 监听数据变化 -->
<template>
  <div class="home">
    <input type="text" v-model="msg">
```

```
    <h3>{{msg}}</h3>
  </div>
</template>

<script>
export default {
  name: 'Home',
  data(){
    return {
      msg: "你好世界"
    }
  },
  watch: {
    msg(newVal, oldVal){
      //newVal 为修改后的数据,oldVal 为 "你好世界"
      console.log(newVal, oldVal)
    }
  }
}
</script>
```

### 5.3.2 立即监听

监听只运行在数据修改之后,若想在数据刚开始绑定时立刻执行监听操作,就需要使用"立即监听",代码如下:

```
<!-- 第 5 章 Vue.js 高级语法:立即监听实现 -->
<template>
  <div class="home">
    <input type="text" v-model="msg">
  </div>
</template>

<script>
export default {
  name: "Home",
  data() {
    return {
      msg: "你好"
    };
  },
  watch: {
    msg: {
      handler(val, oldVal) {
        console.log(123)   //在最初绑定时直接打印 123
        //console.log(val, oldVal);
```

```
      },
      immediate: true
    }
  }
};
</script>
```

**注意** immediate 需要搭配 handler 一起使用，其在最初绑定时，调用的函数也就是这个 handler 函数。

### 5.3.3 深度监听

当需要监听一个对象的改变时，普通的 watch()方法无法监听到对象内部属性的改变。只有 data 中的数据才能监听到变化，此时就需要 deep 属性对该对象进行深度监听，代码如下：

```
<!-- 第 5 章 Vue.js 高级语法：深度监听实现 -->
<template>
  <div class="home">
    <h3>{{obj.age}}</h3>
    <button @click="btnClick">按钮</button>
  </div>
</template>
<script>
export default {
  name: 'Home',
  data(){
    return {
      obj: {
        name: "Lucy",
        age: 13
      }
    }
  },
  methods: {
    btnClick(){
      this.obj.age = 33;
    }
  },
  watch: {
    obj: {
      handler(val, oldVal){
        //运行结果: 33 33
        console.log(val.age, oldVal.age)
      },
```

```
      deep: true
    }
  }
}
</script>
```

### 5.3.4 deep 优化

深度监听不仅可以监听一整个对象，也可以直接监听对象中的某个属性的变化，代码如下：

```
<!-- 第5章 Vue.js 高级语法 -->
<template>
  <div class="home">
    <h3>{{obj.age}}</h3>
    <button @click="btnClick">按钮</button>
  </div>
</template>

<script>
export default {
  name: "Home",
  data() {
    return {
      obj: {
        name: "Lucy",
        age: 13
      }
    };
  },
  methods: {
    btnClick() {
      this.obj.age = 33;
    }
  },
  watch: {
    //通过点语法获取对象中的属性，然后转换为字符串，即可完成对深度监听的优化
    "obj.age": {
      handler(val, oldVal) {
        ...do something
      },
      deep: true
    }
  }
};
</script>
```

本节主要讲解 Watch 侦听器，包括普通的监听方式、立即监听与深度监听。若要监听一个对象，则可以直接对该对象进行监听，亦可监听其包含的某个属性。

## 5.4 动态组件与组件缓存

### 5.4.1 动态组件

开发者自定义的组件，除了直接在父级组件引入、注册及调用，还可以使用 component 组件来展示。假设现在有 3 个子组件，代码如下：

```html
<!-- 定义 3 个子组件 -->
<template>
    <div>子组件 A</div>
</template>

<template>
    <div>子组件 B</div>
</template>

<template>
    <div>子组件 C</div>
</template>
```

在父组件中使用 component 组件结合 is 属性来调用组件，代码如下：

```html
<!-- 第 5 章 Vue.js 高级语法：使用 component 组件结合 is 属性来调用组件 -->
<template>
  <div>
     <component is="AComp"></component>
     <component is="BComp"></component>
     <component is="CComp"></component>
  </div>
</template>

<script>
import AComp from "@/components/AComp";
import BComp from "@/components/BComp";
import CComp from "@/components/CComp";
export default {
  components: {
    AComp,
    BComp,
    CComp
  }
};
</script>
```

> 注意 例如 table、select 这种外层 HTML 标签，它们内部的标签（如 tr、option 等）是不能被单独拿出来作为组件放在它们内部的，否则会被作为无效的内容提升到外部，并导致最终渲染结果出错，而使用动态组件的 :is 属性就可以解决这个问题。

### 5.4.2 KeepAlive 缓存组件

#### 1. include 与 Excelude

有时 keep-alive 需要指定哪些组件需要缓存，哪些组件不需要，此时就需要借助 include 或 Excelude 属性来包含或排除指定的组件。

如果想要缓存指定的组件，则可以使用 include，代码如下：

```
<keep-alive :include="['acom', 'bcom']">
    <component :is="mycom"></component>
</keep-alive>
```

如果想要排除部分组件的缓存，则可以使用 Excelude，代码如下：

```
<keep-alive :Excelude="['ccom']">
    <component :is="mycom"></component>
</keep-alive>
```

#### 2. KeepAlive 生命周期

被<keep-alive></keep-alive>嵌套的组件的生命周期包含 activated()和 deactivated()，代码如下：

```
<!-- KeepAlive 生命周期 -->
<script>
export default {
  activated(){
      //当进入当前组件时触发
  },
  deactivated(){
      //当离开当前组件时触发
  }
}
</script>
```

> 注意 activated()在 created()和 mounted()之后调用，并且 deactivated()在离开组件时会触发，无论第几次进入组件，activated()都会触发，但 created()和 mounted()只会触发一次。

本节主要讲解了动态组件的渲染、KeepAlive 缓存组件，包括如何指定与排除组件，以及 KeepAlive 所包含组件的生命周期。

## 5.5　Vue.js 其他高级用法

Vue.js 还有一些常用但相对零散的知识点，本节将为读者一一讲解。

### 1. Mixins

Mixins 就是定义一部分公共的方法或者计算属性，然后混入各个组件中使用，方便管理与统一修改。同一个生命周期，混入对象会比组件先执行。

在 src 下创建 mixins 目录，在 mixins 目录下创建 index.js 文件，然后导出一个包含 created() 生命周期的模块，代码如下：

```javascript
//导出Mixins模块,包含一个created()生命周期
export const MixinsFn = {
    created() {
        console.log("这是Mixins触发的created")
    }
}
```

然后在组件中引入 Mixins 模块，代码如下：

```vue
<!-- 第5章Vue.js高级语法：Mixins模块引入 -->
<template>
  <div>
    <h1>首页</h1>
  </div>
</template>
<script>
import { MixinsFn } from '@/mixins/index.js'
export default {
  created(){
    console.log("这是当前组件触发的created")
  },
  //引用引入的Mixins模块:MixinsFn
  mixins: [MixinsFn]
}
</script>
```

**注意**　Mixins 引入的生命周期将具备更高的优先级，即 Mixins 的 created 会比当前组件的 created 先执行。

### 2. ref

#### 1）ref 属性与 $refs

当在 Vue.js 文件中获取页面里的某个元素（标签或组件）时可以给它绑定 ref 属性，有点类似于给它添加 id 名，代码如下：

```
<!-- 第5章Vue.js高级语法：通过this.$refs获取具有ref属性的标签或组件 -->
```

```
<template>
  <div class="">
    <h3 ref="title">{{msg}}</h3>
    <button @click="btnclick">按钮</button>
  </div>
</template>

<script>
export default {
  data() {
    return {
      msg: "你好"
    };
  },
  methods: {
    btnclick() {
      console.log(this.$refs.title);              //得到 h3 标签
      console.log(this.$refs.title.innerHTML);    //得到 h3 标签的文本
    }
  }
};
</script>
```

**2)$refs 获取子组件数据与方法**

挂载过 ref 属性的子组件,其携带的方法与 data 中的数据都可以被父组件使用$refs 获取,代码如下:

```
<!-- 为子组件挂载 ref 属性 -->
<Child ref="sub" />

//调用子组件 data 中的数据
this.$refs.sub.num

//调用子组件的方法 fn
this.$refs.sub.fn()
```

**3. NextTick**

当 Vue.js 侦听到页面数据发生变化时,Vue.js 将开启一个队列(该队列被 Vue.js 官方称为异步更新队列)。视图需要等队列中所有数据变化完成之后,再统一进行更新。如果想提前获取更新后的数据,就需要借助 NextTick,代码如下:

```
// (1) CallBack 形式
this.$nextTick(()=>{
  //获取带有 ref 属性的某元素的文本内容
  console.log(this.$refs.title.innerHTML);
})
```

```
// (2) Promise 形式
this.$nextTick().then(()=>{
  //获取带有 ref 属性的某元素的文本内容
  console.log(this.$refs.title.innerHTML);
})
```

#### 4. 图片资源引入方式

当 Vue.js 给 img 标签动态地添加路径时，需要在 JS 中引入本地图片，可以使用 CommonJS 规范，或者 ES6 语法，代码如下：

```
//ES6 引入
import ImgSrc1 from "本地图片资源的相对路径"

//CommonJS 引入
var ImgSrc2 = require("本地图片资源的相对路径")
```

#### 5. Transition 过渡动画

CSS 中有个 transition 属性，称为过渡属性。Vue.js 文件中也有个 Transition 组件，可以把过渡效果应用到其包裹的内容上，而不会额外渲染 DOM 元素，也不会出现在可被检查的组件层级中，代码如下：

```
<!-- 第 5 章 Vue.js 高级语法：通过 Transition 组件控制显示/隐藏动画 -->
<button @click="flag = !flag">切换 h3 显示</button>
<transition name="fade">
  <h3 v-show="flag">你好，标题</h3>
</transition>

<style lang="less" scoped>
//入场初始状态及离场最终状态
.fade-enter, .fade-leave-to{
  opacity: 0;
}

//入场与出场过渡态
.fade-enter-active, .fade-leave-active{
  transition: opacity 1s;
}

//入场最终状态及离场初始状态
.fade-enter-to, .fade-leave{
  opacity: 1;
}
</style>
```

Transition 组件的用法如图 5-4 所示，主要在于控制进出场状态，也可以认为这种状态就是样式。

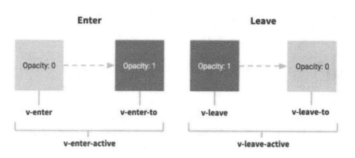

图 5-4　Transition 组件的用法

本节主要讲解了 Mixins 混入、ref 获取元素与调用子组件方法、NextTick 获取更新后的数据、CommonJS 与 ES6 引入图片的方式，以及 Transition 过渡动画。

# 第 6 章

# Vuex

Vuex 是一个专为 Vue.js 应用程序开发的状态管理模式。Vuex 采用集中式存储管理应用的所有组件的状态，并以相应的规则保证状态以一种可预测的方式发生变化。Vuex 已被集成到 Vue.js 的官方调试工具 DevTools Extension (Opens New Window)中，提供了诸如零配置的 time-travel 调试、状态快照引入/导出等高级调试功能。

## 6.1　DevTools

DevTools 的安装很简单，首先打开 Chrome 浏览器的扩展程序界面，如图 6-1 所示。

图 6-1　Chrome 浏览器扩展程序

打开扩展程序界面，如图 6-2 所示，单击左上角打开侧边栏，如图 6-3 所示，单击"打开 Chrome 网上应用店"按钮。

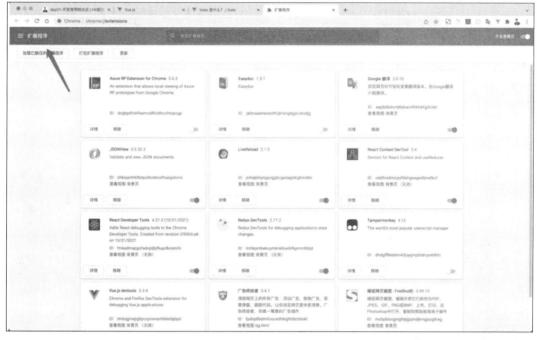

图 6-2　扩展程序界面

图 6-3　扩展程序界面侧边栏

打开 Chrome 网上应用店后，搜索 vue js devtools，按 Enter 键，从搜索结果中选择提供方为"Vue.js 官网"的扩展程序，如图 6-4 所示。

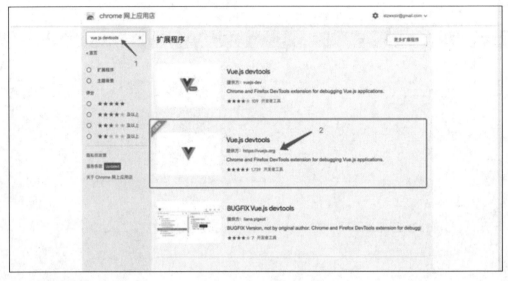

图 6-4　扩展搜索

进入后，单击"添加至 Chrome"按钮，如图 6-5 所示，即可添加此 Chrome 插件。

然后打开 Vue.js 的官网，按 F12 键，即可打开控制台，如果看到选项栏中有 Vue.js 选项，如图 6-6 所示，则代表扩展插件安装成功。

图 6-5　添加至 Chrome

图 6-6　Vue 扩展插件安装成功

> 注意　国内由于网络限速等原因，导致无法打开 Chrome 网上应用店。若无法打开网上应用店安装扩展，建议使用百度搜索 vue-devtools.crx，下载并直接拖曳进扩展程序界面进行安装。

## 6.2　Vuex

### 6.2.1　Vuex 简介与安装

#### 1. Vuex 简介

Vuex 是一个专为 Vue.js 应用程序开发的状态管理模式。它采用集中式存储管理应用的所有组件的状态，并以相应的规则保证状态以一种可预测的方式发生变化。具体流程如图 6-7 所示。Vuex 就是 Vue.js 文件中管理数据状态的一个库，通过创建一个集中的数据存储，供程序中所有组件访问。

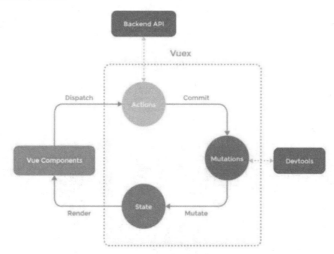

图 6-7　Vuex 流程图

#### 2. Vuex 安装

Vue.js 的脚手架提供了 Vuex 的安装选项，可以直接在创建项目时就勾选，但如果一个项目未曾勾选使用 Vuex，就需要在项目根目录执行安装命令，代码如下：

```
$ npm install vuex --save
```

命令行安装完成后，在项目入口文件中可显式地通过 Vue.use() 来安装 Vuex，代码如下：

```
import Vue from 'vue'
import Vuex from 'vuex'

Vue.use(Vuex)
```

虽然可以手动安装，但如果条件允许，则建议在项目创建时就勾选 Vuex 选项。

## 6.2.2 Vuex 核心概念

Vuex 核心概念有 State、Getters、Mutations、Actions 及 Modules，共 5 个属性。

### 1. State

State 属性用于存放数据，数据以 key:value 的形式存储。在项目根目录下创建 store 文件夹，并在其中创建 index.js 文件作为仓库入口文件，代码如下：

```
//Vuex 数据定义
import Vue from 'vue'
import Vuex from 'vuex'

Vue.use(Vuex)

export default new Vuex.Store({
  state: {
    num: 0//定义了一个 num
  }
})
```

在 State 中定义的数据，可以在任意组件中调用，代码如下：

```
<h3>{{$store.state.num}}</h3>
```

运行结果如图 6-8 所示。

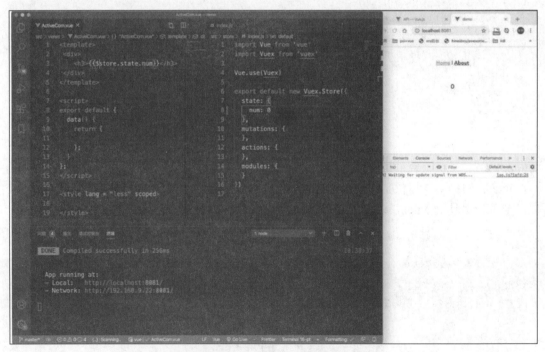

图 6-8　组件调用 Vuex 数据

但在 HTML 中写这么长代码，始终有点难以阅读，因此，可以在 computed 中获取这个值，再传入 HTML 中，代码如下：

```
//以 computed 的形式调用 Vuex 中的数据
computed: {
  num(){
    return this.$store.state.num
  }
}


<!-- 标签中-->
<h3>{{num}}</h3>
```

### 2. Getters

Vuex 中的 getters 类似于 computed 计算属性，getters 的返回值会根据它的依赖被缓存起来，并且只有当它的依赖值发生了改变时才会被重新计算。对 store 文件夹下的 index.js 文件进行修改，代码如下：

```
//第 6 章 Vuex：使用 getters 对 state 中的数据进行计算
import Vue from 'vue'
import Vuex from 'vuex'

Vue.use(Vuex)

export default new Vuex.Store({
  state: {
    num: 2
  },
  getters: {
    //这里的参数 state 可以让开发者快速获取仓库中的数据
    doubleNum(state) {
      return state.num * 2;
    }
  }
})
```

通过计算得到新值在组件中使用，代码如下：

```
<!-- 第 6 章 Vuex：组件中调用 getters 中的数据 -->
<template>
  <div>
    <h3>{{num}}</h3>
  </div>
</template>

<script>
export default {
```

```
  computed: {
    num(){
      return this.$store.getters.doubleNum
    }
  }
};
</script>
```

### 3. Mutations

官网明确指出：更改 Vuex 的 store 中的状态的唯一方法是提交 Mutations，因此，修改 state 中的 num，只能触发 Mutations 中的方法，代码如下：

```
//通过提交 mutations 修改 state 中的数据
import Vue from 'vue'
import Vuex from 'vuex'

Vue.use(Vuex)

export default new Vuex.Store({
  state: {
    num: 2
  },
  mutations: {
    //payload 的专业名称为"载荷"，其实就是个参数
    addNum(state, payload) {
      state.num += payload;
    }
  }
})
```

组件中使用 commit()方法调用 Mutations 中的方法，代码如下：

```
this.$store.commit('addNum', 2);    //第 2 个参数是要传递给 payload 的数据
```

**注意** Mutations 中不允许出现异步操作，它必须是一个同步函数。

### 4. Actions

Actions 类似于 Mutations，不同点在于 Actions 可以包含任意异步操作，但它提交的是 Mutations，而不是直接变更状态，代码如下：

```
//第 6 章 Vuex: actions 中不能直接修改 num，只能调用 Mutations 中的方法
import Vue from 'vue'
import Vuex from 'vuex'

Vue.use(Vuex)

export default new Vuex.Store({
```

```
    state: {
      num: 2
    },
    mutations: {
      addNum(state, payload) {
        state.num += payload;
      }
    },
    actions: {
      //context 是一个对象，包含了 commit 和 state
      AsyncAddNum(context,payload) {
        setTimeout(() => {
          context.commit('addNum', payload)
        }, 1000)
      }
    }
})
```

组件中使用dispatch()方法进行调用，代码如下：

```
this.$store.dispatch('AsyncAddNum', 2);    //第2个参数是传给payload的实参
```

### 5. Modules

由于使用单一状态树，应用的所有状态会集中到一个比较大的对象。当应用变得非常复杂时，store 对象就有可能变得相当臃肿。

为了解决以上问题，Vuex 允许开发者将 store 分割成模块（Module）。每个模块拥有自己的 state、mutations、actions、getters、甚至是嵌套子模块，即从上至下进行同样方式的分割，代码如下：

```
//使用 Modules 来管理多个模块
const moduleA = {
  state: () => ({ ... }),
  mutations: { ... },
  actions: { ... },
  getters: { ... }
}

const moduleB = {
  state: () => ({ ... }),
  mutations: { ... },
  actions: { ... }
}

const store = new Vuex.Store({
  modules: {
    a: moduleA,
    b: moduleB
```

```
    }
})

store.state.a //-> moduleA 的状态
store.state.b //-> moduleB 的状态
```

### 6. 辅助函数

调用 Vuex 中的方法与数据，除了可以直接使用 this.$store.xxx 调用，开发者还可以选择使用数组形式的辅助函数，代码如下：

```
//辅助函数的使用
import {mapState, mapGetters, mapMutations, mapActions} from 'vuex'

//调用 State 中的 num
...mapState(['num'])

//调用 Getters 中的 doubleNum
...mapGetters(['doubleNum'])

//调用 Mutations 中的 addNum
...mapMutations(['addNum'])

//调用 Mutations 中的 AsyncAddNum
...mapActions(['AsyncAddNum'])
```

如果 Vuex 使用了 Modules 来分割模块，则辅助函数调用模块中的数据与方法可以使用对象格式，代码如下：

```
//Vuex 模块化使用辅助函数
...mapState({
  'num': state => state.add.num     //module 中 state 值的获取方法需要写函数形式
})

...mapGetters({
  'doubleNum': 'add/doubleNum'
})

...mapMutations({
  'addNum': 'add/addNum'
})

...mapMutations({
  'AsyncAddNum': 'add/AsyncAddNum'
})
```

本节主要讲解了 Vuex 的核心概念属性与辅助函数，以及 Vuex 常用的 DevTools 浏览器扩展。Vuex 的整个数据操作过程都可以借助 DevTools 来观察数据的变化与视图的更新。

# 第 7 章 路由与请求

## 7.1 路由

前端与后端的路由概念略有不同,后端的路由更倾向于代表接口地址,而前端的路由代表的是页面路径。前端的工程化项目基于路由可以实现 SPA,从而让项目页面的管理更加方便。

> **注意** SPA(Single Page Application)又称单页应用。SPA 是一种特殊的 Web 应用,可加载单个 HTML 页面并在用户与应用程序交互时动态地更新该页面。它的优点主要有用户体验好、良好的前后端分离。

### 7.1.1 Vue Router 简介与安装

#### 1. Vue Router 简介

Vue Router 是 Vue.js 官方的路由管理器。它和 Vue.js 的核心深度集成,让构建单页面应用变得易如反掌,包含的功能如下:

(1)嵌套的路由/视图表。
(2)模块化的基于组件的路由配置。
(3)路由参数、查询、通配符。
(4)基于 Vue.js 过渡系统的视图过渡效果。
(5)细粒度的导航控制。
(6)带有自动激活的 CSS class 的链接。
(7)HTML5 历史模式或哈希模式,在 IE9 中自动降级。
(8)自定义的滚动条行为。

#### 2. Vue Router 安装

Vue 的脚手架提供 Vue Router 的安装选项,可以直接在创建项目时勾选,但如果一个项目未曾勾选使用路由,就需要在项目根目录执行安装命令,代码如下:

```
$ npm install vue-router
```

安装完成后，如果在一个模块化工程中使用它，则必须通过 Vue.use()明确地安装路由功能，代码如下：

```
import Vue from 'vue'
import VueRouter from 'vue-router'

Vue.use(VueRouter)
```

虽然可以手动安装，但如果条件允许，则建议在项目创建时就勾选 Vue Router 选项。

## 7.1.2 路由文件配置

### 1. 路由入口文件配置

假设目前项目中有两个页面（或组件），分别是 Home.vue 与 User.vue。在项目的根目录下创建 router 文件夹，并在该文件夹下创建 index.js 文件，以此作为 router 的入口文件。

**注意** 如果是通过脚手架直接预选了 Vue Router，则以上创建 router 文件夹与 index.js 文件的步骤可以省略，因为 Vue CLI 会自动新建。

路由的入口文件一般用于指定路由路径对应的组件，代码如下：

```
//第 7 章路由与请求：路由文件配置
import Vue from 'vue'
import VueRouter from 'vue-router'
import Home from '@/views/Home.vue'

Vue.use(VueRouter)    //引入路由对象

//定义路由规则
const routes = [
  {
    path: '/',
    name: 'Home',
    component: Home
  },
  {
    path: '/user',
    name: 'User',
    //路由懒加载
    component: () => import(/* webpackChunkName: "user" */ '../views/User.vue')
  }
]

//创建路由实例
const router = new VueRouter({
  mode: 'history',    //Vue 路由只有两种模式，即哈希模式与历史模式
```

```
    base: process.env.BASE_URL,
    route
})

export default router
```

可以看到，路由主要有两种引入组件的方式。一种是使用 import 前置引入；另一种是路由懒加载。

**2．路由懒加载的 3 种方式**

实际上，开发者有 3 种懒加载方式，代码如下：

```
//方式一：结合 Vue 的异步组件和 Webpack 的代码分析
const User = resolve => { require.ensure(['@/views/User.vue'], () =>
{ resolve(require('@/views/User.vue')) }) };

//方式二：AMD 写法
const User = resolve => require(['@/views/User.vue'], resolve);

//方式三：在 ES6 中，可以有更加简单的写法来组织 Vue 异步组件和 Webpack 的代码分割
const Home = () => import(/* webpackChunkName: "user" */ '../views/User.vue')
```

Vue.js 文件中运用 import 的懒加载语句及 Webpack 的魔法注释，在项目进行 Webpack 打包时，对不同模块进行代码分割。当运行了 npm run build 命令之后，在生成的 JS 文件中就能看到以魔法注释定义的 JS 文件名。

### 7.1.3 路由跳转

路由跳转主要有两种方式，一种是 router-link 组件跳转；另一种是编程式导航。

**1．router-link 与 router-view**

使用<router-link></router-link>可以实现路由跳转，代码如下：

```
<!-- 第 7 章路由与请求：router-link 路由跳转 -->
<div id="nav">
  <router-link to="/">Home</router-link> |
  <router-link to="/user">User</router-link>
</div>

<!-- 使用 router-view 来展示路由地址对应的组件 -->
<router-view/>
```

router-view 用于展示路由对应的组件，而 router-link 最终会被渲染为 a 标签，相当于使用 a 标签进行页面跳转。

**2．编程式导航**

所谓编程式导航，即使用 JS 事件的方式进行跳转，代码如下：

```
//页面跳转的完整写法(path)
```

```
this.$router.push({path: '/user'})
```

```
//页面跳转的完整写法(name)
this.$router.push({name: 'User'})
```

```
//页面跳转简写
this.$router.push('/user')
```

除了push()，还有replace()、go()、forward()、back()这几种方法来触发不同情况的跳转。

### 3. 参数携带

参数携带有以下几种方式。

**1）router-link 携带参数**

router-link 可以通过 to 属性跳转路由并携带参数，代码如下：

```
<!-- params 携带参数 -->
<router-link :to="{name:'user',params:{userId: 123}}">User</router-link>
```

```
<!-- 参数携带于网址栏 -->
<router-link to="/user/123">User</router-link>
```

**注意** 若将参数直接携带于网址栏，则参数必须是数字或英文，而且路由配置中必须标明该参数的字段（如/user/:userId?），否则会导致路由无法对应。这种配置中，":"符号代表路径地址中该位置为携带的参数，"?"符号代表该参数可有可无。

**2）path 跳转携带参数**

如果使用 path 跳转携带复杂参数，则只能使用 query 携带，代码如下：

```
this.$router.push({path: '/user', query: {userId: 456}});
```

**3）name 跳转携带参数**

如果使用 name 跳转携带复杂参数，则可以使用 query 携带，也可以使用 params 携带，代码如下：

```
this.$router.push({name: 'User', params: {userId: 456}});
//或
this.$router.push({name: 'User', query: {userId: 456}});
```

以上所有路由跳转所携带的参数，均可以通过 this.$route 在组件中获取对应的路由地址及参数，代码如下：

```
console.log(this.$route);//获取当前路由的所有信息
```

### 7.1.4 导航守卫

Vue Router 提供的导航守卫主要用来监听路由的进入和离开。Vue Router 提供了

beforeEach()和 afterEach()的钩子函数，它们会在路由即将改变前和改变后触发。

### 1. 前置导航守卫

【示例 7-1】 例如一个页面必须登录才能进入，这种情况就可以在路由文件中借助导航守卫实现。

代码如下：

```
//定义全局前置导航守卫
router.beforeEach((to, from, next) => {
//进入购物车页面必须先登录，否则跳转至登录页面
if (to.path == '/cart') {
    //获取token，如果有token，则代表已经登录，如果无token，则提示登录
    let token = localStorage.getItem("token");
    if (!token) {
        alert('先登录');
        next('/login')
        return;
    }
}
  next(); //没有next()方法，导航不会跳转
})
```

### 2. 后置导航守卫

后置导航守卫是进入页面后才触发，因此不需要主动调用 next()，代码如下：

```
router.afterEach( route => {
  console.log(route)
})
```

### 3. 导航元信息 meta

每个路由可以自带一些信息，方法是使用 meta 属性，代码如下：

```
{
  path: '/user',
  component: () => import ('@/views/User.vue'),
  meta: {
      //将keepAlive定义为true，表示该组件希望被缓存
      keepAlive: true
  }
}
```

通过$route.meta 可以获取当前路由的元信息。

### 4. KeepAlive 缓存页面

结合路由元信息 meta 中定义的 keepAlive 属性，可以指定缓存的页面，代码如下：

```
<!-- 有 keepAlive 元信息的就进行缓存 -->
<keep-alive>
  <router-view v-if="$route.meta.keepAlive" />
```

```
</keep-alive>
<router-view v-if="!$route.meta.keepAlive" />
```

本节主要讲解了路由的安装、配置、跳转、传参、守卫、元信息及页面缓存。

## 7.2 请求

前后端之间的交互，通常是前端通过 HTTP 方法向后端发送请求，后端返回数据。当前工程化项目中，最流行的是使用 Axios 库进行数据请求，因此，本节主要讲解 Axios 在 Vue 中的请求实现与封装。

### 1. RESTful 风格接口

RESTful 风格的 API 是一种软件架构风格，是设计风格而不是标准，只是提供了一组设计原则和约束条件。它主要用于客户端和服务器端交互类的软件。基于这个风格设计的软件可以更简洁，更有层次，更易于实现缓存等机制。

在 RESTful 风格中，用户请求的 URL 使用同一个 URL 而用请求方式 get、post、delete、put 等对请求的处理方法进行区分，这样可以在前后端分离式的开发中使前端开发人员不会对请求的资源地址产生混淆和大量的检查方法名的麻烦，形成一个统一的接口。

### 2. Axios 简介与安装

Axios 是目前最流行、易用、简洁且高效的 HTTP 库。官网网址为 http://www.axios-js.com/。可以在项目的根目录下进行命令行安装，代码如下：

```
npm install axios
```

### 3. get 与 post 请求

get 与 post 请求是开发中最常见的请求方式，使用方式也非常简单，代码如下：

```
import axios from 'axios'

axios.get(url[, config])
axios.post(url[, data[, config]])

//get 请求的具体格式
axios
  .get(url, { params: {} })
  .then(res=>{})
  .catch(err=>{})
//post 请求的具体格式
axios
  .post('/user', {})
  .then(res=>{})
  .catch(err=>{})
```

> **注意** 此处不难发现，对于 get 请求，官方要求携带的参数需要嵌套在 params 属性中，而 post 却不需要。

#### 4. 请求与响应拦截器封装

在实际开发中，对于 RESTful 风格的接口，一般会进行请求与响应封装。项目中可以在根目录下创建 request 文件夹，并在该文件夹下创建 request.js 文件，用于书写拦截器，代码如下：

```javascript
//第7章路由与请求：请求与响应拦截器封装
import axios from 'axios'
const instance = axios.create({
  baseURL: '填写请求的地址', //这里 baseURL 不是驼峰式，URL 必须大写
  timeout: 1000 //
});

//添加请求拦截器
instance.interceptors.request.use(function(config) {
  //在发送请求之前做些什么
  return config;
}, function(error) {
  //对请求错误做些什么
  return Promise.reject(error);
});

//添加响应拦截器
instance.interceptors.response.use(function(response) {
  //对响应数据做点什么
  return response;
}, function(error) {
  //对响应错误做点什么
  return Promise.reject(error);
});

export default instance;
```

使用方式也需要经过一次处理，在 request 文件夹下创建 api.js 文件，并按需导出请求方法，代码如下：

```javascript
import request from './request'

//get 请求示例
export const GetDataAPI = (params) => request.get('/getdata', {params});

//post 请求示例
export const PostDataAPI = (data) => request.post('/postdata', params);
```

在组件中引入并调用方法，由于 Axios 底层是经过 Promise 封装的，因此成功与失败的回调可以使用 then 与 catch 接收，代码如下：

```
//引入 API 并调用，实现发送请求
import {GetDataAPI,PostDataAPI} from '@/request/api'

//get 请求
GetDataAPI({num: 1}).then(res => {
  console.log(res);
}).catch(err=>{
  //do something
})

//post 请求
PostDataAPI({num: 1}).then(res=>{
  console.log(res)
}).catch(err=>{
  //do something
})
```

本章主要讲解了 Vue.js 的第三方请求库 Axios，并介绍了如何使用它进行 get 与 post 请求，同时针对 RESTful 风格的接口进行 request 封装。

# 第 8 章 Vue.js 3.0 新增语法

Vue.js 在 3.0 版本（下文简称 Vue 3）中还新增了一些语法，本书也进行补充介绍。由于网上对 Vue 3 的官网复制问题严重，导致开发者经常误入其他个人网站，Vue 3 官网网址为 https://v3.cn.vuejs.org/。

## 8.1 Vue.js 3.0 起步

### 1. 脚手架安装

对于 Vue 3，开发者需要使用 npm 上可用的 Vue CLI v4.5 作为@vue/cli@next，命令如下：

```
$ npm install -g @vue/cli@next
```

然后在 Vue.js 项目运行更新，代码如下：

```
$ vue upgrade --next
```

### 2. 创建项目

创建项目的方式有两种，第 1 种创建方式是直接通过脚手架创建。本书使用命令行创建一个项目，名称为 demo，命令如下：

```
$ vue create demo
```

直接选择 Vue 3 Preview，按 Enter 键，如图 8-1 所示。

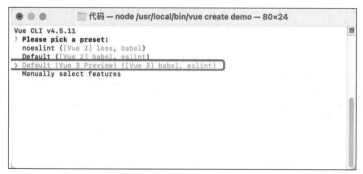

图 8-1 脚手架选择 Vue 3 Preview

项目创建成功后可以运行项目，命令如下：

```
#进入目录
$ cd demo

#运行项目
$ npm run serve
```

第 2 种创建方式是使用 Vite 创建，命令如下：

```
$ npm init vite-app demo
#相当于
$ npx create-vite-app demo

#安装后执行
$ npm install

#运行项目
$ npm run dev
```

创建项目并运行，会发现执行 npm run dev 命令后是秒开项目的，运行速度极快。

## 8.2 Vue.js 3.0 新增语法

### 8.2.1 Composition API

Composition API（组合式 API）相当于 React 中的 Hook，此处使用 Vue 2 的方式实现一个累加，同时使用 Composition API 实现一个累加效果，再进行对比。

Vue 2 实现累加，代码如下：

```
<!-- 第 8 章 Vue.js 3.0 新增语法：Vue 2 实现累加 -->
<template>
  <h2>{{count}}</h2>
  <button @click="btnClick">累加</button>
</template>

<script>
export default {
    data(){
        return {
            count: 0
        }
    },
    methods: {
        btnClick(){
```

```
        this.count++;
      }
    }
  }
</script>
```

以上 Vue 2 实现累加的代码虽然没问题，但如果以后开发者想把这个组件中的 count 字段与 btnClick()方法单独进行管理，那就比较麻烦了，因为 count 和 btnClick()不在同一种方法内，很难抽离。

#### 1. setup()

setup()有以下特性：

（1）setup()函数是处于生命周期函数 beforeCreate 和 Created 两个钩子函数之间的函数，也就是在 setup()函数中是无法使用 data 和 methods 中的数据和方法的。

（2）setup()函数是 Composition API 的入口。

（3）在 setup()函数中定义的变量和方法最后都需要返回，否则无法在模板中使用。

（4）由于开发者不能在 setup()函数中使用 data 和 methods，所以 Vue.js 为了避免开发者错误地使用它们，直接将 setup()函数中的 this 修改成了 undefined。

（5）setup()函数只能是同步的，而不能是异步的。

#### 2. ref()

若使用 Vue 3 实现累加效果，就需要用到 ref()这个 Hook，代码如下：

```
<!-- 第8章 Vue.js 3.0新增语法：ref()实现累加 -->
<template>
  <h2>{{count}}</h2>
  <button @click="btnClick">累加</button>
</template>

<script>
import {ref} from 'vue'//引入ref

export default{
  name: '',
  setup(){
    let count = ref(0);//通过ref定义响应式的数据
    let btnClick = () => {
      count.value++;
    }
    return { count, btnClick }
  }
}
</script>
```

此时如果想单独管理这个累加效果，则可以做代码抽离，代码如下：

```
<!-- 第8章 Vue.js 3.0 新增语法：累加功能单独管理 -->
<template>
  <h2>{{count}}</h2>
  <button @click="btnClick">累加</button>
</template>

<script>
import {ref} from 'vue'

export default{
  name: '',
  setup(){
  //返回 clickCountFn 函数调用的结果
    return clickCountFn();
  //如果后期还想同时返回其他数据，则可以将 clickCountFn()的返回结果展开
    //return {...clickCountFn(), 其他数据}
  }
}

//将原本 setup 中的代码抽离，封装成一个 clickCountFn 函数
function clickCountFn(){
  let count = ref(0);
    let btnClick = () => {
      count.value++;
    }
    return { count, btnClick }
}
</script>
```

### 3. reactive()

reactive()函数和 ref()的作用非常接近，但是它的参数是一个对象，开发者可以在对象中定义其方法，而通过这个形式，就不需要再对其进行 .value 调用了。

同样地，使用 reactive()也可以实现累加，代码如下：

```
<!-- 第8章 Vue.js 3.0 新增语法：reactive()的使用方式 -->
<template>
  <h2>{{count.num}}</h2>
  <button @click="btnClick">累加</button>
</template>

<script>
import {reactive} from 'vue'//引入 reactive

export default{
  name: '',
  setup(){
    //注意，reactive 中必须传入一个对象，而 ref 可以是一个值
```

```
    let count = reactive({num: 0});//通过ref定义响应式的对象数据
    let btnClick = () => {
      count.num++;
    }
    return {count, btnClick}
    //
  }
}
</script>
```

使用reactive生成的对象与使用ref生成的值都是响应式的。

**注意** 不能使用return {...count, btnClick}扩展count对象返回，只能使用count.num返回，这是因为reactive返回的对象本质上已经是一个Proxy对象，使用扩展运算符会导致响应式无效。

### 4. toRefs()

toRefs()能将响应式对象转换为普通对象，其中结果对象的每个property都指向原始对象相应property的ref。开发者可以借助toRefs()实现reactive的扩展运算符返回，代码如下：

```
<!-- 第8章 Vue.js 3.0新增语法：toRefs()实现reactive的扩展运算符返回 -->
<template>
  <h2>{{num}}</h2>
  <button @click="btnClick">累加</button>
</template>

<script>
import {reactive, toRefs} from 'vue'

export default{
  name: '',
  setup(){
    let count = reactive({num: 0});

    //转换为普通对象后，就可以在return中扩展返回
    const countAsRefs = toRefs(count);

    let btnClick = () => {
      count.num++;
    }
    return {...countAsRefs, btnClick}
  }
}
</script>
```

## 8.2.2 Provide 与 Inject

在 Vue.js 文件中实现跨级组件的传值，通常有以下 3 种方式：
（1）无限次的 props 父传子。
（2）中央事件总线。
（3）Vuex。

而实际上，这 3 种方式都不是最好的解决方案，Vue.js 提供了 Provide 与 Inject 来解决这个问题。Provide 与 Inject 相比于 props 的好处在于：如果组件嵌套较多，则 props 需要一级一级地往下传递，后期很难维护。Provide+Inject 相当于跨级组件传值，例如孙子组件想使用父组件 setup 中的 num 值，就不用一级一级地往下传，直接在孙子组件使用即可，代码如下：

```
//第8章 Vue.js 3.0新增语法：父组件通过Provide赋值msg
import {provide,ref} from 'vue'

export default {
  setup(){
    let msg = ref("你好世界");
    provide("msg", msg)
    return {msg}
  }
}

<!-- 子组件通过Inject接收msg -->
<template>
  <h2>{{msg}}</h2>
</template>

<script>
import {inject} from 'vue'

export default{
  setup(){
    const msg = inject("msg", "hello world")
    return {msg}
  }
}
</script>
```

## 8.2.3 Teleport

在 Vue 2 中，当想要将子节点渲染到存在于父组件以外的 DOM 节点时，需要通过第三方库 portal-vue (opens new window)去实现，而在 Vue 3 中，Teleport（传送门）是一种能够将指定的模板移动到 DOM 中 Vue App 之外的其他位置的技术。

在项目的根目录下找到 index.html 文件,并在#app 同级的地方创建#text,代码如下:

```
<div id="app"></div>
<div id="test"></div>
```

然后找到 HelloWorld.vue 文件,书写两个标题,将这两个标题分别传送到#app 和#test 中,代码如下:

```
<!-- Teleport 的实现 -->
<template>
  <p>这段内容渲染在#app 中</p>
  <teleport to="#app">
    <h2>这个标题渲染在#app 中</h2>
  </teleport>
  <teleport to="#test">
    <h2>这个标题渲染在#test 中</h2>
  </teleport>
</template>
```

最终渲染的效果如图 8-2 所示,在一个组件中,开发者可以将不同的内容传送到页面中不同的区块,这就是 Vue 3 的传送门。

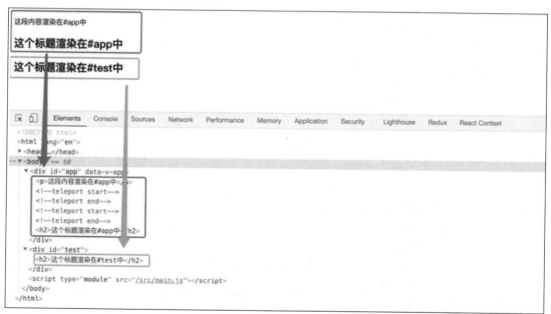

图 8-2　Teleport 传送门

### 8.2.4　Suspense

Suspense 组件用于在等待某个异步组件解析时显示后备内容。示例 8-1 中一个子组件的内容需要经过一次异步获取再返回，此时父组件需要在子组件内容返回之前显示"正在加载中"。

首先，子组件实现异步返回数据，代码如下：

```
<!-- 第8章Vue.js 3.0新增语法：子组件实现异步返回数据 -->
<template>
  <h2>{{msg}}</h2>
</template>

<script>
import {ref} from 'vue'
export default{
  async setup(){
    let msg = ref("Hello World");
    msg = await new Promise(resolve=>{
      setTimeout(()=>{
        resolve("你好世界")
      }, 2000)//2s后才返回数据
    })
    return {msg}
  }
}
</script>
```

然后，父组件需要引入子组件，并在Suspense中使用template（结合#default和#fallback）实现后备内容显示，代码如下：

```
<!-- Suspense使用方式 -->
<template>
  <Suspense>
    <template #default>
      <HelloWorld />
    </template>
    <template #fallback>
      <p>正在加载中...</p>
    </template>
  </Suspense>
</template>
```

注意 #default定义了要写入的内容，#fallback定义了要预显示的内容。

### 8.2.5 Fragment

Vue 2中每个template标签下必须有根元素，意味着开发者必须在组件的最外层套上一个无用的标签，从而导致HTML层级过多。

Vue 3中有了Fragment，就可以直接在template下不写根元素，代码如下：

```
<!-- Fragment -->
<template>
  <Fragment>
    <h2>{{msg}}</h2>
    <p>你好世界</p>
  </Fragment>
</template>

<!-- 相当于直接不写Fragment -->
<template>
    <h2>{{msg}}</h2>
    <p>你好世界</p>
</template>
```

> **注意** Fragment虽然会显示控制台中的HTML，但不会被当作标签渲染。另外，在Suspense中的template不支持Fragment。

### 8.2.6 TreeShaking

TreeShaking 是一个术语，指的是在打包构建过程中移除没有被引用的代码，这些代码可以称为 Dead Code。这个概念最早在基于 ES6 的打包工具 Rollup 中被提出，后来被引入 Webpack 中。TreeShaking 比较依赖于 ES6 模块系统的静态结构特性，例如 import 和 export。

举个例子：Vue 2.x 中常使用 Vue.nextTick(()=>{}) 来预操作 DOM，但有时开发者并不用 nextTick，例如改用别的方式来代替（如 setTimeout），那么项目打包时，Vue.js 全局的 nextTick 就成为多余的代码，从而使项目打包体积变大。

在 Vue 3 中，官方团队重构了所有全局 API 的组织方式，让所有的 API 都支持了 TreeShaking，所以如果在 Vue 3 中还想使用全局的 nextTick，则不能直接调用，而是需要引入，代码如下：

```
import { nextTick } from 'vue';

nextTick(() => {
    //和 DOM 有关的一些操作
});
```

如果开发者在 Vue 3 中不引入而直接调用 Vue.nextTick()，就会得到一个报错：undefined is not a function。Vue.js 官方也给出了 Vue 2.x 中受此更改影响的全局 API：Vue.nextTick、Vue.observable（用 Vue.reactive 替换）、Vue.version、Vue.compile（仅全构建）、Vue.set（仅兼容构建）、Vue.delete（仅兼容构建）。

### 8.2.7　Performance 提升

Vue 3 相对于 Vue 2 来讲性能提高 1.2～1.5 倍，主要原因如下。

**1. diff 方法优化**

Vue 2 中的虚拟 DOM 进行全量对比，而 Vue 3 新增了静态标记（PatchFlag），只比对带有 PF 的节点，并且通过 Flag 的信息得知当前节点要比对的具体内容。

**2. 静态提升**

Vue 2 中无论元素是否参与更新，每次都会重新创建，然后渲染；Vue 3 中对于不参与更新的元素会做静态提升，只会被创建一次，在渲染时直接复用即可。

**3. cacheHandlers 事件侦听器缓存**

在默认情况下 onClick 会被视为动态绑定，所以每次都会去追踪它的变化，但是因为是同一个函数，所以没有追踪变化，直接缓存起来复用即可。

**4. ssr 渲染**

当有大量静态的内容时，这些内容会被当作纯字符串推进一个 buffer 里面，即使存在动态绑定，也会通过模板插值嵌入进去。这样会比通过虚拟 DOM 渲染快很多。

当静态内容大到一定量级时，会用_createStaticVNode 方法在客户端生成一个静态节点，这些静态节点不需要创建对象，然后根据对象渲染。

### 8.2.8　生命周期

Vue 3 对 Vue 2 的生命周期做了一些修改与补充，Vue 2 中的 beforeDestroy() 与 destroyed() 分别被替换为 beforeUnmount() 与 unMounted()。另外新增了 errorCaptured()、renderTracked() 与 renderTriggered()，分别用于捕获错误、跟踪虚拟 DOM 重新渲染，以及虚拟 DOM 重新渲染被触发时的检测。

本节主要讲解了 Vue 3 新增的语法与使用规范，随着 Vue.js 官网的更新，关于 Vue 3 的知识点也在不断增多，建议读者多留意官网关于 Vue 3 内容的更新。

# 第 9 章 项目一：Vue.js 2.0 全家桶+Element 开发后台管理系统

## 9.1 创建项目与添加 Element 模块

**1. 创建项目**

在指定目录下创建项目，命令如下：

```
> vue create v2-management
```

**2. 添加 Element 模块**

创建完项目后，将 Element 引入项目中，命令如下：

```
> vue add element
```

选择按需引入和语言：

```
> ?How do you want to import Element? Import on demand
?Choose the locale you want to load zh-CN
```

安装之后，可以看到 src 目录下多了 plugins 目录，里面是 element.js 文件，包括 Element 组件的引入和注册。

在 main.js 文件中多了一行引入 element.js 文件的代码：

```
import './plugins/element.js'
```

还在 App.vue 文件中使用了 el-button 组件（e-button 组件是 Element 的按钮组件），代码如下：

```
<!-- 第 9 章 Vue.js 2.0 全家桶+Element 开发后台管理系统：使用了 el-button 组件 -->
<template>
  <div id="app">
    <img src="./assets/logo.png">
    <div>
      <p>
        If Element is successfully added to this project, you'll see an
```

```
        <code v-text="'<el-button>'"></code>
        below111222
      </p>
      <!-- el-button 组件 -->
      <el-button>el-button</el-button>
    </div>
  </div>
</template>
```

## 9.2 项目初始化

开始项目之前,需要对项目进行样式初始化的准备工作,把项目目录整理成所需要的目录结构。

删除 App.vue 文件中的 Helloworld.vue 组件的引入和注册,删除 App.vue 文件中的默认样式,代码如下:

```
<!-- 第 9 章 Vue.js 2.0 全家桶+Element 开发后台管理系统:项目初始化 -->
<template>
  <div id="app">
    <router-view></router-view>
  </div>
</template>
```

然后到 Home.vue 文件中测试,代码如下:

```
<!-- 第 9 章 Vue.js 2.0 全家桶+Element 开发后台管理系统:项目初始化 -->
<template>
  <div class="home">
    <el-button>el-按钮</el-button>
    <HelloWorld msg="Welcome to Your Vue.js App"/>
  </div>
</template>
```

在 views 下新建 login 目录,并在 login 目录下新建 Login.vue 组件,然后在路由表配置登录路由,代码如下:

```
<!-- 第 9 章 Vue.js 2.0 全家桶+Element 开发后台管理系统:项目初始化 -->
{
  path: '/login',
  component: () => import(/* webpackChunkName: "login" */ '../views/login/Login.vue')
}
```

在浏览器中输入 http://localhost:8080/login 便可查看效果。

## 9.3 登录组件的初步引入及使用

Element 表单组件的网址为 https://element.eleme.cn/#/zh-CN/component/form,在 Login.vue 文件中添加,代码如下:

```
<!-- 第 9 章 Vue.js 2.0 全家桶+Element 开发后台管理系统:登录组件的初步引入及使用 -->
<el-form :model="ruleForm" status-icon :rules="rules" ref="ruleForm"
label-width="100px" class="demo-ruleForm">
  <el-form-item label="密码" prop="pass">
    <el-input type="password" v-model="ruleForm.pass" autocomplete="off">
  </el-input>
  </el-form-item>
  <el-form-item label="确认密码" prop="checkPass">
    <el-input type="password" v-model="ruleForm.checkPass" autocomplete=
"off">
  </el-input>
  </el-form-item>
  <el-form-item label="年龄" prop="age">
    <el-input v-model.number="ruleForm.age"></el-input>
  </el-form-item>
  <el-form-item>
    <el-button type="primary" @click="submitForm('ruleForm')">提交
</el-button>
    <el-button @click="resetForm('ruleForm')">重置</el-button>
  </el-form-item>
</el-form>

...
<script>
export default {
  data () {
    return {
      ruleForm:{},
      rules:{}
    }
  }
}
</script>
```

在 src/plugins/element.js 文件中注册相关组件,代码如下:

```
<!-- 第 9 章 Vue.js 2.0 全家桶+Element 开发后台管理系统:登录组件的初步引入及使用 -->
import Vue from 'vue'
import { Button,Form,FormItem,Input } from 'element-ui'

Vue.use(Button)
```

```
Vue.use(Form)
Vue.use(FormItem)
Vue.use(Input)
```

## 9.4 登录组件的初步完善

### 9.4.1 登录页面

先初始化项目样式，安装 reset-css 扩展，代码如下：

```
npm i reset-css
```

在 src/main.js 文件中添加 import "reset-css"，代码如下：

```
<!-- 第9章 Vue.js 2.0全家桶+Element 开发后台管理系统：登录组件的初步完善 -->
import Vue from 'vue'
import "reset-css"
import App from './App.vue'
import router from './router'
import store from './store'
import './plugins/element.js'
...
```

在 Login.vue 组件中，实现登录模块结构，代码如下：

```
<!-- 第9章 Vue.js 2.0全家桶+Element 开发后台管理系统：登录组件的初步完善 -->
<template>
    <div class="login-page">
        <div class="login-box">
            <h1>e 店邦 O2O 平台</h1>
            <el-form :model="ruleForm" status-icon :rules="rules" ref="ruleForm" label-width="100px" class="demo-ruleForm">
                <el-form-item label="用户名" prop="pass">
                    <el-input type="text" v-model="ruleForm.pass" autocomplete="off"></el-input>
                </el-form-item>
                <el-form-item :label="'密\xa0\xa0\xa0\xa0码'" prop="checkPass">
                    <el-input type="password" v-model="ruleForm.checkPass" autocomplete="off"></el-input>
                </el-form-item>
                <el-form-item label="验证码" prop="age">
                    <div class="captcha-box">
                        <el-input v-model.number="ruleForm.checkCaptcha"></el-input>
                        <img src="" alt="" height="40">
                    </div>
                </el-form-item>
```

```html
            <el-form-item>
                <el-button class="login-btn" type="primary" @click="submitForm('ruleForm')">登录</el-button>
            </el-form-item>
        </el-form>
    </div>
  </div>
</template>

<script>
export default {
    data () {
        return {
            ruleForm:{},
            rules:{}
        }
    }
}
</script>

<style lang = "less" scoped>
.login-page{
    width: 100%;
    height:100%;
    background: url("../../assets/loginBg.jpg") center top no-repeat;
    position: relative;
    .login-box{
        h1{
            text-align: center;
            font-size: 20px;
            margin-bottom: 30px;
            text-indent: 40px;
        }
        width: 400px;
        background-color: #fff;
        padding: 20px 40px 10px 0;
        border-radius: 20px;
        position: absolute;
        left: 50%;
        top: 50%;
        transform: translate(-50%, -50%);
        .login-btn{
            width: 100%;
        }
    }
}
```

```css
.captcha-box{
    display: flex;
    img{
        margin-left: 20px;
    }
}
</style>
```

### 9.4.2　覆盖 Element UI 样式的正确写法

在 Login.vue 组件中，通过::v-deep 实现 Element UI 样式覆盖，代码如下：

```html
<!-- 第 9 章 Vue.js 2.0 全家桶+Element 开发后台管理系统：登录组件的初步完善 -->
<el-form-item id="login-btn-box">
    <el-button class="login-btn" type="primary" @click="submitForm('ruleForm')">登录</el-button>
</el-form-item>
```

```css
/* 覆盖 Element UI 样式的正确写法 */
::v-deep #login-btn-box .el-form-item__content{
    margin-left: 40px!important;
}
```

### 9.4.3　书写校验规则

先把模板中的 ruleForm.xxx 改成想要的单词，像 ruleForm.username 这种，代码如下：

```html
<!-- 第 9 章 Vue.js 2.0 全家桶+Element 开发后台管理系统：登录组件的初步完善 -->
<el-form-item label="用户名" prop="username">
    <el-input type="text" v-model="ruleForm.username" autocomplete="off"></el-input>
</el-form-item>
<el-form-item :label="'密\xa0\xa0\xa0\xa0码'" prop="password">
    <el-input type="password" v-model="ruleForm.password" autocomplete="off"></el-input>
</el-form-item>
<el-form-item label="验证码" prop="captchacode">
    <div class="captcha-box">
    <el-input v-model.number="ruleForm.captchacode"></el-input>
</el-form-item>
```

在 Login.vue 组件的 data 中书写校验规则，代码如下：

```javascript
<!-- 第 9 章 Vue.js 2.0 全家桶+Element 开发后台管理系统：登录组件的初步完善 -->
data () {
```

```
        return {
            ruleForm:{},
            rules:{
                username:[
                    {
                        required:true,   //必填项
                        message:"账号不能为空!",  //提示语
                        trigger:"blur"   //触发时机
                    }
                ],
            }
        }
```

### 9.4.4 自定义校验规则

在 Login.vue 组件中，定义自定义校验规则，代码如下：

```
<!-- 第 9 章 Vue.js 2.0 全家桶+Element 开发后台管理系统：登录组件的初步完善 -->
username:[
    {
        required:true,  //必填项
        message:"账号不能为空!",  //提示语
        trigger:"blur"  //触发时机
    },
    //自定义校验方法
    {
        validator:this.validUsername,
        trigger:"blur"
    }
]
...
methods:{
    validUsername(rule, value, callback){
        if(value.length<3||value.length>20){
            callback("用户名必须在 3 到 20 个字符内");
        }else{
            callback();
        }
    }
}
```

### 9.4.5 校验

单击"登录"按钮进行校验，代码如下：

```html
<!-- 第 9 章 Vue.js 2.0 全家桶+Element 开发后台管理系统：登录组件的初步完善 -->
<el-button style="width:100%" type="primary"
@click="submitLogin('ruleForm')">提交</el-button>
...
<script>
methods:{
   ...
   submitLogin(form){
       this.$refs[form].validate((vali)=>{
           console.log(vali);
           if(vali){
              //完成登录功能
           }
       })
   }
}
</script>
```

### 9.4.6 企业级项目验证

在/src 下新建 utils 文件夹，新建 validate.js 文件，定义用户名验证方法，代码如下：

```
<!-- 第 9 章 Vue.js 2.0 全家桶+Element 开发后台管理系统：登录组件的初步完善 -->
export function validUsername(rule, value, callback) {
   if(value.length<3||value.length>20){
      callback("用户名必须在 3 到 20 个字符内");
   }else{
      callback();
   }
}
```

在 Login.vue 组件中，导入用户名验证方法，代码如下：

```
<!-- 第 9 章 Vue.js 2.0 全家桶+Element 开发后台管理系统：登录组件的初步完善 -->
import {validUsername} from "@/utils/validate"
...
username:[
    {
        required:true,  //必填项
        message:"账号不能为空！", //提示语
        trigger:"blur"   //触发时机
    },
    //自定义校验方法
    {
        validator:validUsername,
        trigger:"blur"
    }
],
```

## 9.4.7 验证码图片的获取

先安装 Axios，用于数据请求，命令如下：

```
> yarn add axios
```

引入和使用 Axios 发送登录请求，代码如下：

```
<!-- 第 9 章 Vue.js 2.0 全家桶+Element 开发后台管理系统：登录组件的初步完善 -->
import axios from "axios"
...
created(){
    axios.get("http://xue.cnkdl.cn:23683/prod-api/captchaImage").then(res=>{
        if(res.data.code==200){
            //更新图片
            console.log(res);
            this.captchaSrc="data:image/gif;base64,"+res.data.img
            //保存 UUID
            localStorage.setItem("edb-captcha-uuid",res.data.uuid)
        }

    }).catch(res=>{
        console.log(res);
    })
},
```

在浏览器中可以看到效果。

## 9.5 封装 axios 的拦截器

在 request 目录下新建 request.js 文件，创建 axios 实例，定义请求与响应拦截器，代码如下：

```
<!-- 第 9 章 Vue.js 2.0 全家桶+Element 开发后台管理系统：封装 axios 的拦截器 -->
import axios from "axios"
//创建 axios 实例
const instance = axios.create({
    baseURL: "http://xue.cnkdl.cn:23683",
    timeout: 10000
})

//请求拦截器
instance.interceptors.request.use(config => {
    return config
```

```
}, err => {
    return Promise.reject(err)
})

//响应拦截器
instance.interceptors.response.use(res => {
    return res.data
}, err => {
    return Promise.reject(err)
})
export default instance
```

在 api.js 文件中引入 request，定义登录方法，代码如下：

```
import request from "./request"

//获取验证码请求
export const CaptChaApi = () =>
instance.get("http://kumanxuan1.f3322.net:8360/prod-api/captchaImage");
```

在 Login.vue 文件中更新图片，代码如下：

```
<!-- 第 9 章 Vue.js 2.0 全家桶+Element 开发后台管理系统：封装 axios 的拦截器 -->
import {CaptChaApi} from "@/request/api"
...
created(){
    //完成登录功能
    CaptChaApi().then(res=>{
        if(res.code==200){
            //更新图片
            console.log(res);
            this.captchaSrc="data:image/gif;base64,"+res.img
            //保存 UUID
            localStorage.setItem("edb-captcha-uuid",res.uuid)
        }
    }).catch(res=>{
        console.log(res);
    })
},
```

## 9.6 完善登录模块

**1. 单击验证码图片也要发送请求**

在 Login.vue 组件中，将请求验证码的方法抽取为方法，在 created()方法中调用，以便获取图片验证码，代码如下：

```html
<!-- 第 9 章 Vue.js 2.0全家桶+Element 开发后台管理系统：继续完善登录模块 -->
<img @click="getCaptchaCode()" :src="captchaSrc" alt="" height="40">

...
<script>
  created(){
      this.getCaptchaCode();
  },
  methods:{
      getCaptchaCode(){
          CaptChaApi().then(res=>{
              if(res.code==200){
                  //更新图片
                  console.log(res);
                  this.captchaSrc="data:image/gif;base64,"+res.img
                  //保存 UUID
                  localStorage.setItem("edb-captcha-uuid",res.uuid)
              }

          }).catch(res=>{
              console.log(res);
          })
      },
      ...
  }
}
</script>
```

### 2. 登录请求

在 api.js 文件中封装登录请求方法，代码如下：

```
<!-- 第 9 章 Vue.js 2.0 全家桶+Element 开发后台管理系统：继续完善登录模块 -->
//登录页面
export const LoginApi = (params) => request.post("/prod-api/login",params);
```

在 Login.vue 组件中验证用户信息，执行登录请求，代码如下：

```
<!-- 第 9 章 Vue.js 2.0 全家桶+Element 开发后台管理系统：继续完善登录模块 -->
submitForm(form){
    this.$refs[form].validate((valid) => {
        console.log(valid);
        if (valid) {
            LoginApi({
                username:this.ruleForm.username,
                password:this.ruleForm.password,
                code:this.ruleForm.captchacode,
                uuid:localStorage.getItem("edb-captcha-uuid"),
```

```
        }).then(res=>{
            console.log(res);
        })
    } else {
        console.log('error submit!!');
        return false;
    }
});
}
```

### 3. 发起请求的终极解决方案

在 Login.vue 组件中,将获取验证码设置为异步方法,代码如下:

```
<!-- 第9章 Vue.js 2.0全家桶+Element 开发后台管理系统:继续完善登录模块 -->
methods:{
    async getCaptchaCode(){
        let res = await CaptChaApi()
        if(res.code===200){
            //更新图片
            console.log(res);
            this.captchaSrc="data:image/gif;base64,"+res.img
            //保存 UUID
            localStorage.setItem("edb-captcha-uuid",res.uuid)
        }
        //CaptChaApi().then(res=>{
        //if(res.code==200){
        // (1) 更新图片
        //console.log(res);
        //this.captchaSrc="data:image/gif;base64,"+res.img
        // (2) 保存 UUID
        //localStorage.setItem("edb-captcha-uuid",res.uuid)
        //}
        //}).catch(res=>{
        //console.log(res);
        //})
    },
    submitForm(form){
        this.$refs[form].validate(async (valid) => {
            console.log(valid);
            if (valid) {
                let res = await LoginApi({
                    username:this.ruleForm.username,
                    password:this.ruleForm.password,
                    code:this.ruleForm.captchacode,
                    uuid:localStorage.getItem("edb-captcha-uuid"),
                })
                console.log(res);
```

```
        } else {
            console.log('error submit!!');
            return false;
        }
    });
}
```

## 9.7 错误提示及其统一处理方案

### 1. 错误提示

在 src/plugins/element.js 文件中注册 Message 组件,代码如下:

```
import { Button,Form,FormItem,Input,Message } from 'element-ui'
Vue.prototype.$message = Message;
```

在验证码请求方法中添加信息提示,代码如下:

```
<!-- 第9章 Vue.js 2.0全家桶+Element开发后台管理系统:错误提示及其统一处理方案 -->
async getCaptchaCode(){
    let res = await CaptChaApi()
    if(res.code===200){
        //更新图片
        console.log(res);
        this.captchaSrc="data:image/gif;base64,"+res.img
        //保存 UUID
        localStorage.setItem("edb-captcha-uuid",res.uuid)
    }else{
        this.$message({
            message: res.errmsg || "网络请求错误",
            type: 'error'
        });
    }
},
submitForm(form){
    this.$refs[form].validate(async (valid) => {
        console.log(valid);
        if (valid) {
            ...

            if(res.code===200){
                console.log(res);
            }else{
                this.$message({
                    message: res.errmsg || "网络请求错误",
                    type: 'error'
```

```
                });
            }
        } else {
            console.log('error submit!!');
            return false;
        }
    });
}
```

### 2. 统一在响应拦截器中处理错误提示

在 request.js 文件中将错误提示封装在响应拦截器中,代码如下:

```
<!-- 第 9 章 Vue.js 2.0 全家桶+Element 开发后台管理系统:错误提示及其统一处理方案 -->
import { Message } from 'element-ui'
instance.interceptors.response.use(res => {
    let data = res.data
    if(data.code!==200){
        Message({
            message: data.msg || "网络请求错误",
            type: 'error'
        });
    }
    return data
}, err => {
    return Promise.reject(err)
})
```

在 Login.vue 组件中去掉 if-else,代码如下:

```
<!-- 第 9 章 Vue.js 2.0 全家桶+Element 开发后台管理系统:错误提示及其统一处理方案 -->
...
    async getCaptchaCode(){
        let res = await CaptChaApi()
        if(res.code==200){
            //更新图片
            console.log(res);
            this.captchaSrc="data:image/gif;base64,"+res.img
            //保存 UUID
            localStorage.setItem("edb-captcha-uuid",res.uuid)
        }
    },
    submitForm(form){
        this.$refs[form].validate(async (valid) => {
            console.log(valid);
            if (valid) {
                let res = await LoginApi({
```

```
            username:this.ruleForm.username,
            password:this.ruleForm.password,
            code:this.ruleForm.captchacode,
            uuid:localStorage.getItem("edb-captcha-uuid"),
        });

        if(res.code==200){
            console.log(res);
        }


    } else {
      console.log('error submit!!');
      return false;
    }
  });
}
```

## 9.8 登录成功后跳转到首页

在 Login.vue 组件中，实现登录成功后跳转到首页，代码如下：

```
<!-- 第 9 章 Vue.js 2.0 全家桶+Element 开发后台管理系统：登录成功后跳转到首页 -->
submitForm(form){
    this.$refs[form].validate(async (valid) => {
        console.log(valid);
        if (valid) {
            let res = await LoginApi({
                username:this.ruleForm.username,
                password:this.ruleForm.password,
                code:this.ruleForm.captchacode,
                uuid:localStorage.getItem("edb-captcha-uuid"),
            });

            if(res.code==200){
                //登录成功后，保存 token
                localStorage.setItem("edb-authorization-token", res.token);
                //清除 UUID
                localStorage.removeItem("edb-captcha-uuid");
                //提示登录成功
                this.$message({message:"登录成功！",type:"success"});
                //跳转到首页
                this.$router.push("/");
            }
```

```
        } else {
            console.log('error submit!!');
            return false;
        }
    });
}
```

在 views 中新建 layout 目录,在 homepage 目录下新建 MainLayout.vue 组件。
在路由表中添加 mainlayout 路由,代码如下:

```
<!-- 第 9 章 Vue.js 2.0 全家桶+Element 开发后台管理系统:登录成功后跳转到首页 -->
{
    path: '/',
    name: 'mainlayout',
    component: () => import(/* webpackChunkName: "mainlayout" */ '../views/layout/MainLayout.vue')
}
```

## 9.9 经典三栏布局解决方案

首页经典三栏布局分为 HeaderView、NavBar 和 ContentView 组件。
layout 文件夹的安排如下:

```
<!-- 第 9 章 Vue.js 2.0 全家桶+Element 开发后台管理系统:经典三栏布局解决方案 -->
----views
    |----layout
        |----ContentView.vue
        |----HeaderView.vue
        |----MainLayout.vue
        |----NavBar.vue
```

在 MainLayout.vue 组件中,实现基本结构与样式,代码如下:

```
<!-- 第 9 章 Vue.js 2.0 全家桶+Element 开发后台管理系统:经典三栏布局解决方案 -->
<template>
    <div class="layout">
        <NavBar></NavBar>
        <div class="layout-right">
            <HeaderView></HeaderView>
            <ContentView></ContentView>
        </div>
    </div>
</template>
<script>
import HeaderView from "./HeaderView.vue";
import ContentView from "./ContentView.vue";
```

```
import NavBar from "./NavBar.vue";
export default {
    data () {
        return {
        }
    },
    components:{
        NavBar,ContentView,HeaderView
    }
}
</script>
<style lang = "less" scoped>
.layout{
    height: 100%;
    display: flex;
    .layout-right{
        flex:1
    }
}
</style>
```

在 NavBar.vue 文件中，实现导航栏组件的基本结构与样式，代码如下：

```
<!-- 第 9 章 Vue.js 2.0 全家桶+Element 开发后台管理系统：经典三栏布局解决方案 -->
<template>
    <div class="nav-bar">
    </div>
</template>

<script>
</script>

<style lang = "less" scoped>
  .nav-bar{
    width: 220px;
    height: 100%;
    background-color: #304156;
    box-shadow: 5px 0px 5px #ccc;
  }
</style>
```

在 HeaderView.vue 组件中，实现头部模块的基本结构与样式，代码如下：

```
<!-- 第 9 章 Vue.js 2.0 全家桶+Element 开发后台管理系统：经典三栏布局解决方案 -->
<template>
    <header>
        头部
    </header>
```

```
</template>

<style lang = "less" scoped>
    header{
        height: 84px;
        box-shadow: 0 5px 5px #eee;
    }
</style>
```

在 ContentView.vue 组件中，实现内容模块的基本结构与样式，代码如下：

```
<!-- 第 9 章 Vue.js 2.0 全家桶+Element 开发后台管理系统：经典三栏布局解决方案 -->
<template>
    <div class="layout-content">
        内容
    </div>
</template>
</script>

<style lang = "less" scoped>
.layout-content{
    padding: 20px;
}
</style>
```

如果样式上出现侧边栏被右边头部的阴影挡住的情况，则只需在 NavBar 组件中设置样式 position: relative;即可解决此问题。

## 9.10　书写路由守卫

后台管理系统登录常见的两个登录逻辑：
（1）如果访问登录页面，并且有 token，则跳转到首页。
（2）如果访问的不是登录页面，并且没有 token，则跳转到登录页面。
在 src/router/index.js 文件中添加路由守卫，代码如下：

```
<!-- 第 9 章 Vue.js 2.0 全家桶+Element 开发后台管理系统：书写路由守卫 -->
//路由守卫，导航守卫——全局前置钩子函数
router.beforeEach(async (to,from,next)=>{

    let token = localStorage.getItem("edb-authorization-token");

    //如果访问登录页面，并且有 token，则跳转到首页
    if(to.path=="/login"&& token){
        router.push("/")
        return
    }
```

```
        //如果访问的不是登录页面，并且没有token，则跳转到登录页面
    if(to.path!="/login"&& !token){
      router.push("/login")
       return
    }

    next() //放行
})
```

## 9.11 手写菜单栏

### 9.11.1 折叠"菜单"按钮的初步规划

在 HeaderView.vue 组件中实现"菜单"按钮的初步规划，代码如下：

```
<!-- 第9章 Vue.js 2.0 全家桶+Element 开发后台管理系统：手写菜单栏 -->
<template>
    <header>
        <el-button icon="el-icon-s-unfold" v-show="isShow" @click="isShow=!isShow"></el-button><el-button icon="el-icon-s-fold" v-show="!isShow" @click="isShow=!isShow"></el-button>
    </header>
</template>

<script>
export default {
    data () {
        return {
            isShow:true
        }
    }
}
</script>

<style lang = "less" scoped>
    header{
        height: 84px;
        box-shadow: 0 5px 5px #eee;
        .el-button{
            width: 50px;
            height: 50px;
            font-size: 24px;
            padding: 12px 0;
            margin: 0;
            border: 0 none;
```

```
      }
    }
</style>
```

## 9.11.2  菜单展开和折叠状态的展示

Element UI 中菜单组件的网址为 https://element.eleme.cn/#/zh-CN/component/menu#ce-lan。在 NavBar.vue 文件中添加属性 collapse，代码如下：

```
<!-- 第 9 章 Vue.js 2.0 全家桶+Element 开发后台管理系统：手写菜单栏 -->
<template>
    <div class="nav-bar">
      <!-- @open="handleOpen"
    @close="handleClose"

    unique-opened 是否只保持一个子菜单的展开
     -->
      <el-menu
    default-active="2"
    class="el-menu-vertical-demo"
    background-color="#304156"
    text-color="#fff"
    active-text-color="#ffd04b"
    :unique-opened="true"
    :collapse="isNavCollapse">
    <el-submenu index="1">
      <template slot="title">
        <i class="el-icon-location"></i>
        <span>导航一</span>
      </template>
      <el-menu-item index="1-1">选项 1</el-menu-item>
      <el-menu-item index="1-2">选项 2</el-menu-item>
    </el-submenu>
  </el-menu>
  </div>
</template>

<script>
export default {
  data () {
    return {
      isNavCollapse:false      //设置为 true，表示折叠
    }
  }
}
```

```
</script>

<style lang = "less" scoped>
  .nav-bar{
    ...
    .el-menu {
      border:0 none;
    }
  }
</style>
```

### 9.11.3　是否折叠导航栏

只需单击 HeaderView.vue 组件中的"菜单"按钮，修改这个值。将来 Content 可能需要用到这个值，所以把它放在 Vuex 中管理。

在 store 目录下创建 navCollapse 目录，然后在 navCollapse 目录下创建 index.js 文件，添加导航栏是否折叠属性，代码如下：

```
<!-- 第 9 章 Vue.js 2.0 全家桶+Element 开发后台管理系统：手写菜单栏 -->
export default{
    namespaced:true,
    state: {
        isNavCollapse:true
    },
    mutations: {
    },
    actions: {
    },
}
```

在 store/index.js 文件中引入，代码如下：

```
<!-- 第 9 章 Vue.js 2.0 全家桶+Element 开发后台管理系统：手写菜单栏 -->
...
import navCollapse from './navCollapse'
...
export default new Vuex.Store({
  modules: {
    navCollapse
  }
})
```

在 NavBar.vue 组件中实现状态获取，代码如下：

```
<!-- 第 9 章 Vue.js 2.0 全家桶+Element 开发后台管理系统：手写菜单栏 -->
<el-menu
         default-active="2"
```

```
            class="el-menu-vertical-demo"
            background-color="#545c64"
            text-color="#fff"
            active-text-color="#ffd04b"
            :unique-opened="true"
            :collapse="isNavCollapse"
        >
...
<script>
import {mapState} from "vuex"
export default {
    data () {
        return {
        }
    },
    computed:{
        ...mapState({
            isNavCollapse:state=>state.navCollapse.isNavCollapse
        })
    }
}
</script>
```

此时单击按钮则可以看到菜单是否收缩受 Vuex 中值的影响。

### 9.11.4 修改 Vuex 中 isNavCollapse 的值

在 store/navCollapse/index.js 文件中，添加并修改 isNavCollapse()方法，代码如下：

```
<!-- 第 9 章 Vue.js 2.0 全家桶+Element 开发后台管理系统：手写菜单栏 -->
    mutations: {
        changeCollapseVal(state){
            //直接取反
            state.isNavCollapse = !state.isNavCollapse
        }
    },
```

在 HeaderView.vue 组件中，修改 isNavCollapse()的状态，代码如下：

```
<!-- 第 9 章 Vue.js 2.0 全家桶+Element 开发后台管理系统：手写菜单栏 -->
<template>
    <header>
        <div class="header-top">
            <el-button icon="el-icon-s-unfold" v-show="isNavCollapse" @click="changeCollapseVal"></el-button><el-button icon="el-icon-s-fold" v-show="!isNavCollapse" @click="changeCollapseVal"></el-button>
        </div>
    </header>
```

```
</template>

<script>
import {mapState,mapMutations} from "vuex"
export default {
    data () {
        return {
        }
    },
    computed:{
        ...mapState({
            isNavCollapse:state=>state.navCollapse.isNavCollapse
        })
    },
    methods:{
        ...mapMutations({
            "changeCollapseVal":"navCollapse/changeCollapseVal"
        })
    }
}
</script>
```

## 9.11.5 菜单栏折叠卡顿的问题

如果出现菜单栏折叠卡顿现象，则可以在 NavBar.vue 组件中给 el-menu 添加 collapse-transition="false"属性解决此问题。

## 9.11.6 折叠过渡效果的实现

在 NavBar.vue 组件中实现折叠过渡效果，代码如下：

```
<!-- 第 9 章 Vue.js 2.0 全家桶+Element 开发后台管理系统：手写菜单栏 -->
<div class="nav-bar" :class={isCollapse:isNavCollapse}>
</div>
...
<style>
.nav-bar{
    ...
    transition: all .3s;
    &.isCollapse{
        width: 64px;
    }
}
</style>
```

### 9.11.7 补充 Logo 和标题

在 NavBar.vue 组件中添加 Logo 和标题，代码如下：

```html
<!-- 第 9 章 Vue.js 2.0 全家桶+Element 开发后台管理系统：手写菜单栏 -->
<el-menu
    ...
    >
    <h1 class="main-logo">
      <img :src="logoSrc" alt="" width="32">
        <span v-show="!isNavCollapse">通用后台管理系统</span>
    </h1>
<script>
import logo from "@/assets/image/logo.png"

export default {
    data () {
       return {
           logoSrc:logo,
       }
    },
</script>
```

折叠时的样式，代码如下：

```css
.main-logo{
    ...
    position: relative;
    ...
    span{
      ...
      min-width: 130px;
      display: inline-block;
      position: absolute;
      left: 50px;
      top: 16px;
    };
}
```

### 9.11.8 定义初始数据导航

在 NavBar.vue 组件中定义初始数据，代码如下：

```
<!-- 第 9 章 Vue.js 2.0 全家桶+Element 开发后台管理系统：手写菜单栏 -->
data () {
    return {
```

```
            menuData: [//导航栏初始数据
                {
                    title: "首页",
                },
                {
                    title: "客户管理",
                    children: [
                        { title: "客户档案"},
                        { title: "拜访记录"},
                    ],
                },
                {
                    title: "休养预约",
                    children: [
                        { title: "预约信息"},
                        { title: "服务项"},
                        { title: "结算单"},
                    ],
                },
                {
                    title: "流程管理",
                    children: [
                        { title: "审核流程定义"},
                    ],
                },
            ],
        }
    },
```

在 NavBar.vue 组件的模板中，添加子项判断，代码如下：

```
<!-- 第9章 Vue.js 2.0全家桶+Element 开发后台管理系统：手写菜单栏 -->
<!-- 配置自动识别是否有子项目，如果有子项目就用 el-submenu 组件，如果没有就用
el-menu-item -->
<div v-for="(item,idx) in menuData" :key="idx">
    <el-submenu :index="idx" v-if="item.children">
        <template slot="title">
<i class="el-icon-location"></i>
<span>{{item.title}}</span>
        </template>
        <el-menu-item :index="sidx+''" v-for="(sitem,sidx) in
item.children" :key="sidx">{{sitem.title}}</el-menu-item>
    </el-submenu>

    <el-menu-item :index="idx" v-else>
        <i class="el-icon-menu"></i>
        <span slot="title">{{item.title}}</span>
```

```
        </el-menu-item>
</div>
```

修改外层并添加了 div 之后折叠时会出现文字不能隐藏的 Bug，代码如下：

```
<!-- 第 9 章 Vue.js 2.0 全家桶+Element 开发后台管理系统：手写菜单栏 -->
/*隐藏文字*/
.el-menu--collapse  .el-submenu__title span{
    display: none;
}
/*隐藏三角符号> */
::v-deep .el-menu--collapse  .el-submenu__title .el-submenu__icon-arrow{
    display: none;
}
```

### 9.11.9 菜单实现路由跳转

在登录并获取 token 后，后端会返回该用户可以访问的路由列表。

在 NavBar.vue 组件中实现路由跳转配置，代码如下：

```
<!-- 第 9 章 Vue.js 2.0 全家桶+Element 开发后台管理系统：手写菜单栏 -->
<el-menu
    ...
    :router="true"
>
    ...
    <div v-for="(item,idx) in menuData" :key="idx">
        <!-- 【注意】对应的 index 都要修改 -->
        <el-submenu :index="item.path" v-if="item.children">
            ...
            <el-menu-item :index="sitem.path" v-for="(sitem,sidx) in item.children" :key="sidx">{{sitem.title}}</el-menu-item>
        </el-submenu>

        <el-menu-item :index="item.path" v-else>
         ...
        </el-menu-item>
    </div>
    ...
<script>
...
data () {
    return {
        menuData: [
            {
                title: "首页",
                path:"/"
```

```
                },
                {
                    title: "客户管理",
                    path:"/customer",
                    children: [
                        { title: "客户档案", path:"/customer"},
                        { title: "拜访记录", path:"/visit"},
                    ],
                },
                {
                    title: "休养预约",
                    path:"/business",
                    children: [
                        { title: "预约信息", path:"/appointment"},
                        { title: "服务项", path:"/service"},
                        { title: "结算单", path:"/statement"},
                    ],
                },
                {
                    title: "流程管理",
                    path:"/flow",
                    children: [
                        { title: "审核流程定义",path:"/definition"},
                    ],
                },
            ]
        },
        ...
</script>
```

在页面中单击菜单栏可以看见页面已经可以发生跳转。

## 9.12 统一处理请求后 code==200 的情况

在 request.js 响应拦截器中，添加响应数据判断，实现 code 不等于 200 的响应统一处理，代码如下：

```
<!-- 第 9 章 Vue.js 2.0 全家桶+Element 开发后台管理系统：统一处理请求后 code==200 的情况 -->
instance.interceptors.response.use(res => {
    let data = res.data
    if(data.code!==200){
        Message({
            message: data.msg || "网络请求错误",
            type: 'error'
```

```
        });
        //TODO：针对性处理返回的不同参数的情况
        //补充 return false，通知组件里面的请求不是 200，在组件中写请求时就可以阻止不
        //是 200 的情况，并且少写 if(res.code==200) 判断
        return false
    }
    return data
}, err => {
    return Promise.reject(err)
})
```

把 Login.vue 组件之前的两个请求中的 if 判断去掉，但是要加上 if(!res)return 来阻止不是 200 的情况，代码如下：

```
<!-- 第 9 章 Vue.js 2.0 全家桶+Element 开发后台管理系统：统一处理请求后 code==200 的情况 -->
async getCaptchaCode(){
    let res = await CaptChaApi()
    if(!res)return;
    //更新图片
    console.log(res);
    this.captchaSrc="data:image/gif;base64,"+res.img
    //保存 UUID
    localStorage.setItem("edb-captcha-uuid",res.uuid)

},
submitForm(form){
    this.$refs[form].validate(async (valid) => {
        console.log(valid);
        if (valid) {
            let res = await LoginApi({
                username:this.ruleForm.username,
                password:this.ruleForm.password,
                code:this.ruleForm.captchacode,
                uuid:localStorage.getItem("edb-captcha-uuid"),
            });

            if(!res)return;
            //登录成功后，保存 token
            localStorage.setItem("edb-authorization-token", res.token);
            //清除 UUID
            localStorage.removeItem("edb-captcha-uuid");
            //提示登录成功
            this.$message({message:"登录成功！",type:"success"});
            //跳转到 home 页面
            this.$router.push("/");
```

```
       }
    ...
```

## 9.13 动态生成菜单栏

因为每个用户的权限不一样,能够访问的路径也不一样,左边的菜单栏展示也不一样,所以需要动态生成菜单栏(menuData 数组),即 menuData 数组要动态生成。

提示:这是每个后台管理系统的通用需求。

### 9.13.1 请求获取用户菜单列表

在用户登录成功后,获取用户可访问菜单列表(权限列表/路由信息)。
在 api.js 文件中添加获取用户菜单列表方法,代码如下:

```
//获取路由信息
export const GetUserRoutersApi = () => request.get("/prod-api/getRouters");
```

在 request.js 请求拦截器中,添加 token 配置,代码如下:

```
<!-- 第 9 章 Vue.js 2.0 全家桶+Element 开发后台管理系统:动态生成菜单栏 -->
const token = localStorage.getItem("edb-authorization-token");
if(token && !config.url.endsWith("/captchaImage")
&& !config.url.endsWith("/login")){
  config.headers["Authorization"] = "Bearer "+ token;
}
```

接着,在 src/router/index.js 路由守卫中发起请求。为什么要在路由守卫中发起用户菜单列表请求呢?因为有可能用户访问的这个页面是没有权限的,例如用户直接在浏览器地址栏输入地址访问,所以在进入这个系统页面之前需进行请求验证,代码如下:

```
<!-- 第 9 章 Vue.js 2.0 全家桶+Element 开发后台管理系统:动态生成菜单栏 -->
import {GetUserRoutersApi} from "@/request/api"
...
//路由守卫,导航守卫——全局前置钩子函数
router.beforeEach(async (to,from,next)=>{

  let token = localStorage.getItem("edb-authorization-token");

  //如果访问登录页面,并且有token,则跳转到首页
  ...
  //如果访问的不是登录页面,并且没有token,则跳转到登录页面
  ...

  //如果没有登录就不获取登录用户的权限列表
  //if(已经登录且权限列表的长度为 0){
```

```
    if (token && store.state.userMenu.userMenuData.length === 1) {
      //发起用户菜单列表的请求
      //这个用户菜单列表的请求要放在路由守卫中,因为用户在进入除/login外的路径时就应该
//获得这个列表了
      let GetUserRoutersApiRes = await GetUserRoutersApi();
      if(!GetUserRoutersApiRes) return;
      console.log(GetUserRoutersApiRes);

    }

    next() //放行
})
```

如果请求成功,则得到如下数组:

```
<!-- 第9章 Vue.js 2.0 全家桶+Element 开发后台管理系统:动态生成菜单栏 -->
{
    "msg": "操作成功",
    "code": 200,//响应代码
    "data": [//响应数据
        {
            "name": "Customer",//名字
            "path": "/customer",//路由路径
            "hidden": false,
            "redirect": "noRedirect",
            "component": "Layout",//组件
            "alwaysShow": true,
            "meta": {//菜单信息
                "title": "客户管理",//标题
                "icon": "people",//图标
                "noCache": false
            },
            "children": [//子菜单
                {
                    "name": "Customer",
                    "path": "customer",
                    "hidden": false,
                    "component": "customer/index",
                    "meta": {
                        "title": "客户档案",
                        "icon": "#",
                        "noCache": false
                    }
                },
                {
                    "name": "Visit",
                    "path": "visit",
```

```json
                "hidden": false,
                "component": "customer/visit/index",
                "meta": {
                    "title": "拜访记录",
                    "icon": "#",
                    "noCache": false
                }
            }
        ]
    },
    {
        "name": "Business",
        "path": "/business",
        "hidden": false,
        "redirect": "noRedirect",
        "component": "Layout",
        "alwaysShow": true,
        "meta": {
            "title": "休养预约",
            "icon": "online",
            "noCache": false
        },
        "children": [
            {
                "name": "Appointment",
                "path": "appointment",
                "hidden": false,
                "component": "business/appointment/index",
                "meta": {
                    "title": "预约信息",
                    "icon": "#",
                    "noCache": false
                }
            },
            {
                "name": "Service",
                "path": "service",
                "hidden": false,
                "component": "business/service/index",
                "meta": {
                    "title": "服务项",
                    "icon": "#",
                    "noCache": false
                }
            },
            {
                "name": "Statement",
```

```
                    "path": "statement",
                    "hidden": false,
                    "component": "business/statement/index",
                    "meta": {
                        "title": "结算单",
                        "icon": "#",
                        "noCache": false
                    }
                }
            ]
        },
        {
            "name": "Flow",
            "path": "/flow",
            "hidden": false,
            "redirect": "noRedirect",
            "component": "Layout",
            "alwaysShow": true,
            "meta": {
                "title": "流程管理",
                "icon": "component",
                "noCache": false
            },
            "children": [
                {
                    "name": "Definition",
                    "path": "definition",
                    "hidden": false,
                    "component": "flow/definition/index",
                    "meta": {
                        "title": "审核流程定义",
                        "icon": "#",
                        "noCache": false
                    }
                }
            ]
        }
    }
```

## 9.13.2 分析思路

上面的 menuData 是管理员账号设置的一个普通用户的菜单。

用户 qdtest1 登录后发送请求获取权限，列表为 9.13.1 节的数据，但是这个格式并不能直接更新 menuData，因为跟需要的 menuData 的格式不一样，所以要把 menuData 按照 9.13.1 节可访问的路径处理成正确的格式。

即当用户 qdtest1 登录后，menuData 应该处理成如下数据，而这个数据就是由权限列表提取和改造而来的），代码如下：

```
<!-- 第 9 章 Vue.js 2.0 全家桶+Element 开发后台管理系统：动态生成菜单栏 -->
[
    {
        title: "首页",
        path:"/"
    },
    {
        title: "客户管理",
        path:"/customer",
        children: [
            { title: "客户档案", path:"/customer/customer"},
            { title: "拜访记录", path:"/customer/visit"},
        ],
    },
    {
        title: "休养预约",
        path:"/business",
        children: [
            { title: "预约信息", path:"/business/appointment"},
            { title: "服务项", path:"/business/service"},
            { title: "结算单", path:"/business/statement"},
        ],
    },
    {
        title: "流程管理",
        path:"/flow",
        children: [
            { title: "审核流程定义",path:"/flow/definition"},
        ],
    },
],
```

## 9.13.3 处理 menuData 数组

在 src/router/index.js 文件中，在路由守卫中处理 menuData，代码如下：

```
<!-- 第 9 章 Vue.js 2.0 全家桶+Element 开发后台管理系统：动态生成菜单栏 -->
//路由守卫，导航守卫 —— 全局前置钩子函数
router.beforeEach(async (to,from,next)=>{
  let token = localStorage.getItem("edb-authorization-token");

  //如果访问登录页面，并且有 token，则跳转到首页
```

```js
    //如果访问的不是登录页面,并且没有token,则跳转到登录页面
    //如果没有登录就不获取登录用户的权限列表
    //if(已经登录且权限列表的长度为0){
    if(token && store.state.userMenuData.menuData.length==0){
        //获取用户的菜单数据
        let GetUserRoutersApiRes = await GetUserRoutersApi();
        if(!GetUserRoutersApiRes)return;
        console.log("GetUserRoutersApiRes 为",GetUserRoutersApiRes);
        let menuDataArr = [{title: "首页",path:"/"}];
        GetUserRoutersApiRes.data.forEach(item => {
          //console.log(item.children);
          if(item.children){//如果有children属性
            //这个this.menuData 应该放在vuex中,在NavBar组件中要获取
            menuDataArr.push({
              title: item.meta.title,
              path:item.path,
              icon:item.meta.icon,
              children:item.children.map(subitem=>({
                  title: subitem.meta.title,
                  path: item.path+"/"+subitem.path
              }))
            })
          }else{
            menuDataArr.push({
              title: item.meta.title,
              path:item.path
            })
          }
        });
        store.commit("userMenuData/changeMenuData",menuDataArr)
        console.log("menuDataArr",menuDataArr);

    }
    next() //放行
})
```

在 Vuex 中部署 menuData,在 store 目录下创建 userMenu 文件夹,创建 index.js 文件,代码如下:

```
<!-- 第 9 章 Vue.js 2.0 全家桶+Element 开发后台管理系统:动态生成菜单栏 -->
export default{
    namespaced:true,
    state:{
      menuData:[]
    },
    mutations:{
        changeMenuData(state,payload){
```

```
        state.menuData=payload
      }
    }
  }
```

从 store/index.js 导入 userMenu,代码如下:

```
<!-- 第 9 章 Vue.js 2.0 全家桶+Element 开发后台管理系统:动态生成菜单栏 -->
import Vue from 'vue'
import Vuex from 'vuex'
import navCollapse from './navCollapse'
import userMenuData from './userMenuData'
Vue.use(Vuex)

export default new Vuex.Store({
  modules: {
    navCollapse,userMenuData
  }
})
```

在 NavBar.vue 组件中获取这个 menuData,代码如下:

```
<!-- 第 9 章 Vue.js 2.0 全家桶+Element 开发后台管理系统:动态生成菜单栏 -->
computed:{
    ...mapState({
        ...
        menuData:state=>state.userMenuData.menuData
    })
},
```

至此,即可切换不同的账号,访问页面能够看见不同的左侧菜单内容。

## 9.14 修改二级菜单栏的样式补充

在 NavBar.vue 组件中实现二级菜单栏的样式补充,代码如下:

```
<!-- 第 9 章 Vue.js 2.0 全家桶+Element 开发后台管理系统:修改二级菜单栏的样式补充 -->
/* 修改二级菜单栏背景色 */
.el-submenu .el-menu-item{
  background-color: rgb(38,52,69)!important;
}
::v-deep .el-submenu__title:hover,.el-menu-item:hover{
  background-color: #444!important;
}
```

## 9.15 图标处理

**1. 学会找解决方案：Vue 项目中如何使用 svg**

图标的展示方案有普通图片、iconfont、svg 等。鉴于 svg 的渲染效率，svg 图标在项目中使用得非常广泛。

在 https://www.iconfont.cn/search/index?searchType=icon&q=people 可获取 svg 图标，在 https://www.jb51.net/article/264663.htm 或者 https://blog.csdn.net/ksjdbdh/article/details/122364113 可展示 Vue 项目中的 svg 图标。

**2. 在项目中动态处理图标**

在 NavBar.vue 组件中获取动态图标，代码如下：

```
<!-- 第 9 章 Vue.js 2.0 全家桶+Element 开发后台管理系统：图标处理 -->
<div v-for="(item,idx) in menuData" :key="idx">
    <el-submenu :index="item.path" v-if="item.children">
        <template slot="title">
            <!-- 换成 svg -->
            <svg-icon :icon-file-name="item.icon" style="margin:0 10px 0 4px"></svg-icon>
            ...
        </template>
        ...
    </el-submenu>

    <el-menu-item :index="item.path" v-else>
        <!-- 换成 svg -->
        <svg-icon icon-file-name="dashboard" style="margin:0 10px 0 4px"></svg-icon>
        ...
    </el-menu-item>
</div>

<script>
async created(){
    ...

    GetUserRoutersApiRes.data.forEach(item => {
        this.menuData.push({
            ...
            //补充icon
            icon: item.meta.icon,
            ...
        })
    });
    console.log(GetUserRoutersApiRes);
```

```
        },
</script>
```

## 9.16  认证失败处理

在 request/request.js 响应拦截器中判断,如果响应的代码为 401,则表示认证失败,跳转到登录页面,代码如下:

```
<!-- 第9章 Vue.js 2.0 全家桶+Element 开发后台管理系统:认证失败处理 -->
if(data.code!==200){
    Message({
        message: data.msg || "网络请求错误",
        type: 'error'
    });
    //TODO:针对性处理返回的不同参数的情况
    if(data.code==401){
        //console.log("跳转到登录页面");
        //跳转到登录页面
        //401 状态码未携带有效的 token
        localStorage.removeItem("edb-authorization-token");
        router.push("/login");
    }
    return false
}
```

## 9.17  配置子路由(内容部分)

在 router/index.js 文件中配置 children 子路由,代码如下:

```
<!-- 第9章 Vue.js 2.0 全家桶+Element 开发后台管理系统:配置子路由(内容部分) -->
{
  path: '/',
  name: 'mainlayout',
  component: () => import(/* webpackChunkName: "mainlayout" */ '../views/layout/MainLayout.vue'),
  redirect:"/home",
  children:[
    {
      path:"/home",
      component: () => import(/* webpackChunkName: "home" */ '../views/HomeView.vue')
    },
    {
```

```
      path:"/customer/customer",
      component: () => import(/* webpackChunkName: "customer" */
'../views/customer/Customer.vue')
    },
    {
      path:"/customer/visit",
      component: () => import(/* webpackChunkName: "visit" */
'../views/customer/Visit.vue')
    },
    {
      path:"/flow/definition",
      component: () => import(/* webpackChunkName: "definition" */
'../views/flow/Definition.vue')
    }
  ]
},
```

在 views 文件夹中新建 customer 和 flow 等文件夹，以及其对应组件，代码如下：

```
<!-- 第 9 章 Vue.js 2.0 全家桶+Element 开发后台管理系统：配置子路由(内容部分) -->
----views
    |----customer
        |----customer.vue
        |----Visit.vue
    |----flow
        |----Definition.vue
```

在/src/views/layout/Content.vue 文件中需要 router-view 组件作占位符，表示渲染路由，代码如下：

```
<div class="layout-content">
    <router-view></router-view>
</div>
```

这样单击时就能跳转了，但是目前路由是固定不变的。

## 9.18 动态添加子路由规则

此时登录 qdtest2 账号，发现 qdtest2 用户本身没有访问/flow/definition 的权限，但是能够在浏览器网址栏上访问 http://localhost:8080/attribute。这就很危险了，用户竟然可以通过网址栏访问他本没有权限访问的路径，所以此时 router/index.js 文件中的子路由也不能固定不变，需要动态添加，即根据用户的权限列表来添加。

**1．整体思路分析**

在路由守卫中，根据 GetUserRoutersApiRes 请求回来的数据，进行遍历处理，得到用户可以访问的子路由。再利用 router.addRoute（父路由名称，单个子路由对象）；把刚才的子路由添加到 mainlayout 路由里面。

### 2. 具体实现

在全局路由守卫中，获取用户可访问的子路由，代码如下：

```
<!-- 第 9 章 Vue.js 2.0 全家桶+Element 开发后台管理系统：动态添加子路由规则 -->
if (token && store.state.userMenu.userMenuData.length === 1) {
    //发起用户菜单列表的请求
    ...
    GetUserRoutersApiRes.data.forEach(item => {
        ...
    });

    //请求完用户菜单数据后
    //【处理用户可访问路由(动态生成路由)】
    var newChildrenRoutes = [{
        path:"/home",
        component: () => import('../views/HomeView.vue')
    }]
    //获取用户可访问的子路由
    GetUserRoutersApiRes.data.forEach(item=>{
        let ret = item.children.map(sitem=>{
            let pp = item.path+"/"+sitem.path[0].toUpperCase()+sitem.path.slice(1);
            return{
                path:item.path+"/"+sitem.path,
                component: () => import('../views/'+pp+'.vue')
            }
        })
        newChildrenRoutes = [...newChildrenRoutes,...ret]
    })
    //console.log(newChildrenRoutes);
    //动态添加路由
    newChildrenRoutes.forEach(item=>{
      router.addRoute("mainlayout",item);
    })
    //console.log(router); //router 的 matcher 的 addRoute 可以看到全部添加进去了

    //【重要】
    //添加完动态路由后要跳转到这个路径
    next(to.path)
    return
}
```

### 3. 根据当前路径高亮选中对应菜单

在 Nav.vue 文件中将 default-active 属性设置为当前路由路径值，代码如下：

```
<el-menu
```

```
    :default-active="$route.path"
...
```

## 9.19 添加路由切换的过渡动画

在 views/layout/ContentView.vue 组件中添加过渡效果，代码如下：

```
<!-- 第 9 章 Vue.js 2.0 全家桶+Element 开发后台管理系统：添加路由切换的过渡动画 -->
<div class="layout-content">
    <transition name="el-fade-in-linear" mode="out-in">
        <router-view></router-view>
    </transition>
</div>
```

也可以自定义实现过渡效果，代码如下：

```
<!-- 第 9 章 Vue.js 2.0 全家桶+Element 开发后台管理系统：添加路由切换的过渡动画 -->
<transition name="fade-transform" mode="out-in">
    <router-view></router-view>
</transition>
...
<style lang = "less" scoped>
...

/* fade-transform */
.fade-transform-leave-active,
.fade-transform-enter-active {
  transition: all .35s;
}

.fade-transform-enter {
  opacity: 0;
  transform: translateX(-10px);
}

.fade-transform-leave-to {
  opacity: 0;
  transform: translateX(10px);
}
</style>
```

## 9.20 面包屑处理

### 9.20.1 渲染和样式初步处理

在文件夹中新建 Crumb.vue 组件，实现面包屑的基本结构与样式，代码如下：

```
<!-- 第 9 章 Vue.js 2.0 全家桶+Element 开发后台管理系统：面包屑处理 -->
<template>
        <div class="crumb">
            <el-breadcrumb separator="/">
                <el-breadcrumb-item :to="{ path: '/' }">首页</el-breadcrumb-item>
                <el-breadcrumb-item>活动详情</el-breadcrumb-item>
            </el-breadcrumb>
        </div>
</template>
<style lang = "less" scoped>
    .crumb{
        float: left;
    }
    .el-breadcrumb{
        line-height: 50px;
    }
</style>
```

在 views/layout/HeaderView.vue 组件中引入 Crumb 组件，代码如下：

```
<!-- 第 9 章 Vue.js 2.0 全家桶+Element 开发后台管理系统：面包屑处理 -->
<template>
        <header>
            <div class="header-top">
                <div class="coll-btn">
                    <el-button icon="el-icon-s-unfold" v-show="isNavCollapse" @click="changeCollapseVal"></el-button><el-button icon="el-icon-s-fold" v-show="!isNavCollapse" @click="changeCollapseVal"></el-button>
                </div>
                <Crumb></Crumb>
                <div class="header-top-right">头像</div>
            </div>
        </header>
</template>

<script>
import Crumb from "@/components/Crumb"
export default {
    components:{
```

```
        Crumb
    },
    ...
}
</script>

<style lang = "less" scoped>
    header{
        ...
        .coll-btn{
            float: left;
        }
        .header-top-right{
            float: right;
        }
    }
</style>
```

## 9.20.2　title 的收集

处理用户可访问路由（动态生成路由），代码如下：

```
<!-- 第 9 章 Vue.js 2.0 全家桶+Element 开发后台管理系统：面包屑处理 -->
    var newChildrenRoutes = [{
        path:"/home",
        //添加 title 数组，用于处理标题数组
        meta:{title:["首页"]},
        component: () => import('../views/HomeView.vue')
    }]

    GetUserRoutersApiRes.data.forEach(item=>{
        let ret = item.children.map(sitem=>{
          let pp = item.path+"/"+sitem.path[0].toUpperCase()+sitem.path.slice(1);
            return{
                path:item.path+"/"+sitem.path,
                //[!!!]
                meta:{title:[item.meta.title,sitem.meta.title]},
                component: () => import('../views'+pp+'.vue')
            }
        })
        newChildrenRoutes = [...newChildrenRoutes,...ret]
    });
    console.log(GetUserRoutersApiRes.data);
    console.log(newChildrenRoutes);
```

### 9.20.3　在面包屑组件中展示 title

回到 crumb.vue 组件中，获取路由中的 title 数组并在页面结构中遍历 title 数组，以便动态生成面包屑结构，代码如下：

```html
<!-- 第9章 Vue.js 2.0全家桶+Element 开发后台管理系统：面包屑处理 -->
<template>
    <div class="crumb">
        <el-breadcrumb separator="/">
            <!-- <el-breadcrumb-item :to="{ path: '/' }">首页
</el-breadcrumb-item> -->
            <!-- 通过 v-for 遍历 title 数组 -->
            <el-breadcrumb-item v-for="item,index in title" :key="index">{{item}}</el-breadcrumb-item>
        </el-breadcrumb>
    </div>
</template>
<script>
export default {
    data () {
        return {
            title:[]
        }
    },
    watch:{
        $route:{
            immediate:true,
            handler(){
                //获取路由的 title 数组
                this.title = this.$route.meta.title;
            }
        }
    }
}
</script>
```

### 9.20.4　解决网址栏跳转但视图不更新的情况

在 App.vue 文件中补充监听浏览器的返回按钮的事件并注意事件的销毁，代码如下：

```
<!-- 第9章 Vue.js 2.0全家桶+Element 开发后台管理系统：面包屑处理 -->
mounted() {
    //监听浏览器的返回按钮
    if (window.history && window.history.pushState) {
        history.pushState(null, null, document.url);
        window.addEventListener("popstate", this.onCloseModal, false);
```

```
        }
    },
    destroyed() {
        window.removeEventListener("popstate", this.onCloseModal, false);
    },
    methods: {
        onCloseModal() {
            console.log("onCloseModal",window.location.pathname);
            this.$router.push(window.location.pathname)
        },
    },
```

## 9.21 404 页面的处理

在/src/router/index.js 文件中，在静态路由中添加 404 页面对应的路由，代码如下：

```
{
    path:"*",
    component: () => import('../views/404.vue')
}
```

在 views 中新建 404.vue 组件，实现基本结构，代码如下：

```
<!-- 第9章 Vue.js 2.0 全家桶+Element 开发后台管理系统：404 页面的处理 -->
<template>
    <div>
        <p>找不到页面！正在跳回原来的页面{{pnts}}</p>
    </div>
</template>

<script>
import router from "@/router"

export default {
    data () {
        return {
            pnts:""
        }
    },
    created(){
        let t = setInterval(()=>{
            this.pnts+=".."
            if(this.pnts==".........."){
                clearInterval(t)
            }
```

```
        },300)
        setTimeout(()=>{
            router.push("/")
        },2000);
    }
}
</script>

<style lang = "less" scoped>
    p{
        padding: 100px;
    }
</style>
```

## 9.22 删除 token

手动打开 Application 面板删除 token，再按 Backspace 键来到登录页面，登录之后显示的是上一个账号的菜单，这是浏览器的本身特性造成的，所以进入登录页面需要清除 Vuex 中的用户菜单数据解决这个 Bug。

在 Login.vue 组件中，在进入登录页面时清空用户菜单数据，代码如下：

```
<!-- 第 9 章 Vue.js 2.0 全家桶+Element 开发后台管理系统：手动打开 Application 面板删除 token -->
    created(){
        ...
        //来到登录页面时清空用户菜单数据
        this.changeMenuData([])
    },
    methods:{
        ...mapMutations({
            "changeMenuData":"userMenuData/changeMenuData"
        }),
        ...
```

## 9.23 用户信息处理

### 9.23.1 登录成功获取用户信息

在/src/request/api.js 文件中，添加登录成功获取用户信息的方法，代码如下：

```
<!-- 第 9 章 Vue.js 2.0 全家桶+Element 开发后台管理系统：用户信息处理 -->
//获取用户信息
export const GetUserInfoApi = () => request.get("/prod-api/getInfo");
```

在 /src/store 中新建 userInfo 文件夹，新建 index.js 文件，代码如下：

```
<!-- 第 9 章 Vue.js 2.0 全家桶+Element 开发后台管理系统：用户信息处理 -->
import { GetUserInfoApi } from "@/request/api"
export default{
    namespaced:true,
    state: {
        //定义用户信息数据
        userInfo:JSON.parse(localStorage.getItem("edb-userInfo"))||{
            permissions:null,
            roles:null,
            user:null
        }
    },
    mutations: {
        //切换用户信息
        changeUserInfo(state,payload){
            state.userInfo={...payload}
            localStorage.setItem("edb-userInfo",JSON.stringify(state.userInfo))
        }
    },
    actions: {
      //异步获取用户信息
      async asyncchangeUserInfo({commit}){
            //执行获取用户信息操作
            let GetUserInfoApiRes = await GetUserInfoApi();
            console.log(GetUserInfoApiRes);
            if(!GetUserInfoApiRes){return}
            //调用用户信息设置
            commit("changeUserInfo",{
                permissions:GetUserInfoApiRes.permissions,
                roles:GetUserInfoApiRes.roles,
                user:GetUserInfoApiRes.user
            })
        }
    },
}
```

登录成功后获取用户信息，在 Login.vue 组件中的代码如下：

```
//发起用户信息请求
this.asyncchangeUserInfo()
```

在 utils 中新建 baseurls，设置图片服务器地址，代码如下：

```
export const IMG_BASEURL = "http://xue.cnkdl.cn:23683"
```

在 HeaderView 组件中，通过图片服务器地址+具体图片地址，渲染用户头像，代码如下：

```html
<!-- 第9章 Vue.js 2.0全家桶+Element开发后台管理系统：用户信息处理 -->
<template>
    <header>
        <div class="header-top">
            ...
            <div class="header-top-right">
                <div class="avatar-box">
                    <img class="avatar" :src="IMG_BASEURL+(this.userInfo.user.avatar||'/prod-api/profile/avatar/2022/10/10/blob_20221010200353A001.jpeg')" width=40 alt="">
                    <i class="iconfont el-icon-caret-bottom"></i>
                </div>
            </div>
        </div>
    </header>
</template>

<script>
import {IMG_BASEURL} from "@/utils/baseurls"
    ...
export default {
    data () {
        return {
            IMG_BASEURL
        }
    },
    components:{
        Crumb
    },
    computed:{
        ...mapState({
            ...
            userInfo:state=>state.userInfo.userInfo
        })
    }
}
</script>

<style lang = "less" scoped>
    ...
    .avatar-box{
        margin-right: 20px;
        cursor: pointer;
        .avatar{
            margin-top: 5px;
            border-radius: 10px;
        }
    }
```

```
        }
</style>
```

## 9.23.2  下拉菜单及退出登录

UI 组件库的网址为 https://element.eleme.cn/#/zh-CN/component/dropdown#ji-chu-yong-fa，代码如下：

```
<!-- 第 9 章 Vue.js 2.0 全家桶+Element 开发后台管理系统：用户信息处理 -->
<div class="header-top-right">
    <!--单击每项都会触发 command,在时间函数中匹配 el-dropdown-item 的 command 属性，
    这样才能知道用户单击了哪一项，也就可以做对应的事情了-->
    <el-dropdown @command="handleCommand">
        <div class="avatar-box">
            ...
        </div>
        <el-dropdown-menu slot="dropdown">
            <el-dropdown-item command="logout">退出登录</el-dropdown-item>
            <!-- <el-dropdown-item>狮子头</el-dropdown-item>
            <el-dropdown-item>螺蛳粉</el-dropdown-item>
            <el-dropdown-item disabled>双皮奶</el-dropdown-item>
            <el-dropdown-item divided>蚵仔煎</el-dropdown-item> -->
        </el-dropdown-menu>
    </el-dropdown>
</div>

<script>
    ...
    methods:{
        ...mapMutations({
            changeCollapseVal:"navCollapse/changeCollapseVal",
        }),
        handleCommand(command){
            if(command=="logout"){
                localStorage.removeItem("edb-authorization-token");
                localStorage.removeItem("edb-userInfo");
                this.$router.push("/login")
            }

        }
    }
</script>
```

## 9.24 标签栏处理

标签栏最终效果如图 9-1 所示。

图 9-1 标签栏最终效果

【示例 9-1】每个标签上的文字用于展示当前所在页面的二级标题。在默认情况下（刷新页面）仅展示首页和当前所在页面的标签。如果当前就是首页，则只展示首页标签。当前所在路由拥有当前样式，即前面的白色圈圈和背景颜色。单击 ⟳ 可以跳转至当前展示信息。单击 ✕ 可以关闭该标签，并跳转到最后一个标签的路由。首页标签一直存在且不可关闭，标签栏管理菜单如图 9-2 所示。

在标签上右击可以展示标签管理菜单，完成每项单击的对应操作。

图 9-2 标签栏管理菜单效果

### 9.24.1 初步布局

在 components 文件夹下新建 Tags 组件，实现基本结构，代码如下：

```
<!-- 第 9 章 Vue.js 2.0 全家桶+Element 开发后台管理系统：标签栏处理 -->
<template>
    <div class="tags">
        <el-tag class="tag" effect="dark" size="medium" closable><i></i>中等标签</el-tag>
        <el-tag class="tag" effect="plain" size="medium" closable><i></i>中等标签</el-tag>
        <el-tag class="tag" effect="plain" size="medium" closable><i></i>中等标签</el-tag>
    </div>
</template>
<style lang = "less" scoped>
/* 外层样式 */
.tags{
    padding-left: 20px;
    padding-top: 4px;
    /* 单个 tag 样式 */
    .tag{
```

```
            cursor: pointer;
        margin-left: 5px;
        font-size:13px;
        i{
                background: #fff;
                display: inline-block;
                width: 8px;
                height: 8px;
                border-radius: 50%;
                position: relative;
                left:-4px;
                margin-right: 2px;

            }
        }
    }
</style>
```

在 HeaderViews.vue 组件中的代码如下：

```
<div class="header-bottom">
    <Tags></Tags>
</div>
```

## 9.24.2 组织 tags 数组

添加 tags 数组，实现动态渲染，代码如下：

```
<!-- 第 9 章 Vue.js 2.0 全家桶+Element 开发后台管理系统：标签栏处理 -->
<template>
    <div class="tags">
        <el-tag class="tag" effect="dark" size="medium" closable
            v-for="item in tags" :key="item.path"
        ><i></i>{{item.title}}</el-tag>
        <!-- <el-tag class="tag" effect="plain" size="medium" closable><i></i>中等标签</el-tag> -->
    </div>
</template>

<script>
export default {
    data () {
        return {
            tags:[
                {
                    title: "首页",
                    path:"/home",
```

```
            }
        ]
    }
},
watch:{
    $route:{
        immediate:true,
        handler(val,oldval){
            //只要路由一变化,就检查这个新的路由是否在 tags 数组里,如果不在就往里面
            //push
            const bool = this.tags.find(item=>{
                return item.path==val.path
            });
            if(!bool){
                this.tags.push({
                    path:val.path,
                    title:this.$route.meta.title[this.$route.meta.title.length-1]
                });
            }
        }
    }
}
</script>
```

## 9.24.3　当前样式的处理

每项都用 isActive 属性来控制,代码如下:

```
<!-- 第 9 章 Vue.js 2.0 全家桶+Element 开发后台管理系统:标签栏处理 -->
<el-tag class="tag" size="medium" closable
    v-for="item in tags" :key="item.path"

    :effect="item.isActive?'dark':'plain'"
    :class="{active:item.isActive}"

><i></i>{{item.title}}</el-tag>
<script>
    data () {
        return {
            tags:[
                {
                    title: "首页",
                    path:"/home",
                    isActive:false
```

```
                    }
                ]
            }
        },
        watch:{
            $route:{
                immediate:true,
                handler(val,oldval){
                    //只要路由一变化,就检查新的路由是否在tags数组里,如果不在就往里面push
                    let i = null;   //用来保存存在这个路径的元素的下标
                    var bool = this.tags.find((item,index)=>{
                        //保存存在这个路径的元素的下标
                        if(item.path==val.path){i=index}
                        return item.path==val.path
                    });
                    //如果没有在里面就push,并设置样式
                    if(!bool){
                        this.tags.push({
                            path:val.path,
                            title:this.$route.meta.title[this.$route.meta.title.length-1]
                        });
                        //给最后一个设置当前样式
                        this.tags = this.tags.map(item=>({...item,isActive:false}));
                        this.tags[this.tags.length-1].isActive=true;
                    }else{
                        //如果在里面,则给出当前样式即可
                        this.tags = this.tags.map(sitem=>({...sitem,isActive:false}));
                        this.tags[i].isActive=true;
                    }
                }
            }
        },
</script>
<style lang = "less" scoped>
.tags{
    ...
    .tag{
        i{
            ...
            display: none;
        }
        ...
    }
    .active i{
        display: inline-block;
    }
}
```

## 9.24.4 跳转处理

给标签添加单击事件,实现单击标签页时能够跳转路由,代码如下:

```html
<!-- 第9章 Vue.js 2.0全家桶+Element开发后台管理系统:标签栏处理 -->
<el-tag class="tag" :effect="item.isActive?'dark':'plain'" size="medium" closable
        :class="{active:item.isActive}"
        v-for="item,index in tags" :key="item.path"
        @click="clickTag(item.path)"
    ><i></i>{{item.title}}</el-tag>

<script>
export default {
    ...
    methods:{
        clickTag(path){
            //跳转页面
            this.$router.push(path)
        }
    }
}
</script>
```

## 9.24.5 删除标签

给标签绑定 close 事件,实现删除标签功能,代码如下:

```html
<!-- 第9章 Vue.js 2.0全家桶+Element开发后台管理系统:标签栏处理 -->
<el-tag ...
        @close="closeTag(index)"
        :disable-transitions="true"
    >
<script>
export default {
    ...
    methods:{
        ...
        closeTag(i){
         //删除当前项
            this.tags.splice(i,1);
         //跳转到最后一个
            this.$router.push(this.tags[this.tags.length-1]);
        }
    }
```

## 9.24.6 右击出现快捷菜单

在 tags 组件中书写右击事件，实现右击出现快捷菜单功能，代码如下：

```
<!-- 第 9 章 Vue.js 2.0 全家桶+Element 开发后台管理系统：标签栏处理 -->
<div class="tags-item"  @contextmenu="rightClick($event)" v-for="item,
index in tags" :key="item.path">
   ...
</div>
<script>
   data () {
      return {
         tags:[
            {
                title: "首页",
                path:"/home",
                isActive:false,
            }
         ],
         isShowTagMenu:false,
         menuX:0,
         menuY:0,
      }
   },
   //在 methods 中定义函数
   methods:{
      rightClick(e,i){
         this.isShowTagMenu=true
         console.log("右击",e.clientX,e.clientY);
         this.menuY=e.clientY
         this.menuX=e.clientX
         //阻止右击时浏览器默认的菜单出现
         window.event.returnValue=false;
         return false
      },
   }
</script>
```

## 9.24.7 菜单项现实逻辑的控制

在 tags 组件中，调用时要传单击的 clickIndex 和 tags 的总长度，代码如下：

```
<!-- 第 9 章 Vue.js 2.0 全家桶+Element 开发后台管理系统：标签栏处理 -->
```

```html
<div class="tags-item" @contextmenu="rightClick($event,index)" v-for="item,
index in tags" :key="item.path">
    ...
    </div>
<TagMenu
    ...
    :clickIndex="clickIndex"
    :tagsLength="tags.length"
    ></TagMenu>
<script>
    data () {
      return {
        ...
        clickIndex:0
      }
    },
    methods:{
      rightClick(e,i){
        ...
        this.clickIndex=i;
        ...
        return false
      },
    }
</script>
```

在 TagMenu 组件中，实现标签显示/隐藏控制功能，代码如下：

```html
<!-- 第9章 Vue.js 2.0全家桶+Element 开发后台管理系统：标签栏处理 -->
<li v-for="item,index in tmenu" :key="index" v-show="isShowLi(index,
clickIndex)">
    <i :class="item.icon"></i>
    {{item.text}}
</li>
<script>
    props:["menuX","menuY","clickIndex","tagsLength"],
    methods:{
      isShowLi(i,clickIndex){
        //如果用户单击的是第 1 个标签，则第 2 个至第 4 个要隐藏
        if(clickIndex==0){
            return ![1,3].includes(i)
        }
        //如果用户单击的是第 2 个标签，则要看一看是不是最后一个
        if(clickIndex==1){
            if(clickIndex==this.tagsLength-1){
                return ![3,4].includes(i)
```

```
            }else{
                return ![3].includes(i)
            }
        }
        //如果用户单击的是最后一个标签,则前面所有标签都要隐藏
        if(clickIndex==this.tagsLength-1){
            return ![4].includes(i)
        }
        return true

    }
}
</script>
```

## 9.24.8 静动态路由的区分

在 views 中创建 Profile.vue 组件,实现基本结构,代码如下:

```
<!-- 第 9 章 Vue.js 2.0 全家桶+Element 开发后台管理系统:标签栏处理 -->
<template>
    <div>
        个人中心页面
    </div>
</template>

<script>
export default {
    data () {
        return {
        }
    }
}
</script>
<style lang = "less" scoped>

</style>
```

**注意** 个人中心的添加,如果只是在默认路由表 routes 里面添加,则面包屑和 tags 栏会有问题。原因是面包屑和 tags 栏都是根据动态路由监听来显示的。

应该在路由守卫中添加路由,代码如下:

```
<!-- 第 9 章 Vue.js 2.0 全家桶+Element 开发后台管理系统:标签栏处理 -->
//【处理用户可访问路由(动态生成路由)】
var newChildrenRoutes = [{
    path:"/home",
    meta:{title:["首页"]},
```

```
    component: () => import('../views/HomeView.vue'),
},{
    path: '/profile',
    meta: {title:["个人中心"]},
    component: () => import('../views/Profile.vue')
}]
```

在 HeaderView 组件中,添加个人中心选项,代码如下:

```
<!-- 第9章 Vue.js 2.0全家桶+Element 开发后台管理系统:标签栏处理 -->
<el-dropdown-menu slot="dropdown">
    <el-dropdown-item command="profile">个人中心</el-dropdown-item>
    <el-dropdown-item command="logout">退出登录</el-dropdown-item>
</el-dropdown-menu>
<script>
    //在 methods 中
handleCommand(command){
    console.log(command);
    if(command=="logout"){
        localStorage.removeItem("edb-authorization-token");
        localStorage.removeItem("edb-userInfo");
        this.$router.push("/login");
        return
    }
    if(command=="profile"){
        this.$router.push("/profile");
        return
    }

}
</script>
```

## 9.24.9 关闭标签栏

在 tags 组件中,实现关闭标签功能,代码如下:

```
<!-- 第9章 Vue.js 2.0全家桶+Element 开发后台管理系统:标签栏处理 -->
mounted() {
    document.addEventListener("click", this.onWinClick);
},
destroyed() {
    document.removeEventListener("click", this.onWinClick);
},
methods:{
    onWinClick(){
        this.isShowTagMenu=false  //关闭标签栏
    },
```

```
        clickTag(path){
            ...
         this.isShowTagMenu=false  //关闭标签栏
        },
        closeTag(i){
            ...
            this.isShowTagMenu=false  //关闭标签栏
        }
}
```

### 9.24.10 根据单击的项目对 tags 进行操作

在 TagsMenu 中，给 data 中的 tmenu 数组补充 id 传到父组件 Tags.vue 文件中操作 tags 会更加方便，代码如下：

```
<!-- 第 9 章 Vue.js 2.0 全家桶+Element 开发后台管理系统：标签栏处理 -->
<li v-for="item,index in tmenu" :key="index" v-show="isShowLi(index,
clickIndex)" @click="hdClick(item.id)">
            ...
        </li>
<script>
    data () {
       return {
          tmenu:[//标签菜单数据
             {
                icon:"el-icon-refresh-right",
                text:"刷新页面",
                id:1    //添加 id
             },
             {
                icon:"el-icon-close",
                text:"关闭当前",
                id:2
             },
             {
                icon:"el-icon-circle-close",
                text:"关闭其他",
                id:3
             },
             {
                icon:"el-icon-back",
                text:"关闭左侧",
                id:4
             },
             {
                icon:"el-icon-right",
```

```
                    text:"关闭右侧",
                    id:5
                },
                {
                    icon:"el-icon-circle-close",
                    text:"全部关闭",
                    id:6
                }
            ]
        }
    },
    methods:{
        hdClick(id){
            this.$emit("fn",id)
        },
        ...
    }
}
</script>
```

在 Tags.vue 文件中，实现所有标签关闭功能，代码如下：

```
<!-- 第 9 章 Vue.js 2.0 全家桶+Element 开发后台管理系统：标签栏处理 -->
<TagMenu
    ...
    @fn="handleTags"
></TagMenu>
<script>
    methods:{
        handleTags(id){
            //关闭全部
            if(id==6){
                this.tags=this.tags.filter((item,index)=>{
                    return this.clickIndex==0?(index==0):(index==0||index==this.clickIndex)
                })
                this.clickTag(this.tags[this.clickIndex].path)
            }
        },
    }
</script>
```

## 9.25　表格处理

在 Api.js 文件中获取客户列表信息，代码如下：

```javascript
//获取客户列表信息
export const GetCustomerListApi = (params) => request.get("/prod-api/customer",{params});
```

在组件中实现获取表格数据、渲染表格和选择表格功能，代码如下：

```html
<!-- 第9章 Vue.js 2.0全家桶+Element开发后台管理系统：表格处理 -->
<template>
    <div>
        <el-table
        ref="multipleTable"
        :data="tableData"

        @selection-change="handleSelectionChange">
        <el-table-column type="selection" width="55" align="center"></el-table-column>
        <el-table-column prop="name" label="客户姓名" align="center"></el-table-column>
        <el-table-column
        align="center"
          label="电话">
          <template slot-scope="{row}">{{ row.phone}}</template>
        </el-table-column>
        <el-table-column
        align="center"
          label="性别">
          <template slot-scope="{row}">{{ row.sex=="1"?"男":(row.sex=="0"?"女":"保密") }}</template>
        </el-table-column>
        <el-table-column
          prop="inputUserName"
          align="center"
          label="录入人">
        </el-table-column>
        <el-table-column
          label="录入时间"
          align="center"
          width="180"
        >
          <template slot-scope="{row}">{{ row.entryTime?new Date(row.entryTime).toLocaleDateString().replaceAll("/","-"):"" }}</template>
        </el-table-column>
        <el-table-column label=" 操作 " align="center" class-name="small-padding fixed-width">
          <template>
            <el-button size="mini" type="text" icon="el-icon-edit"
            >修改
            </el-button>
```

```html
      </template>
    </el-table-column>
  </el-table>
    </div>
</template>

<script>
//导入获取数据方法
import {GetCustomerListApi} from "@/request/api"
export default {
    data () {
        return {
          tableData: [],
          multipleSelection: [],
            pageNum:1,   //当前页数
          pageSize:10, //每页条数
            total:40
        }
    },
    async created(){
        //获取数据
        const GetCustomerListApiRes = await GetCustomerListApi({
            pageNum:this.pageNum,
            pageSize:this.pageSize
        });
        console.log(GetCustomerListApiRes);
        //设置表格数据
        this.tableData = GetCustomerListApiRes.rows
    },
    methods:{
        //选择方法
        toggleSelection(rows) {
            if (rows) {
            rows.forEach(row => {
                this.$refs.multipleTable.toggleRowSelection(row);
            });
            } else {
            this.$refs.multipleTable.clearSelection();
            }
        },
        //在复选框选中数据
        handleSelectionChange(val) {
            this.multipleSelection = val;
        }
    }
}
</script>
```

## 9.26 分页处理

UI 组件文档的网址为 https://element.eleme.cn/#/zh-CN/component/pagination。
实现分页组件的渲染，获取表格总数据，实现分页功能，代码如下：

```html
<!-- 第 9 章 Vue.js 2.0 全家桶+Element 开发后台管理系统：分页处理 -->
<!--盒子居右的正确写法-->
<div style="display:flex;justify-content:flex-end;padding:20px 20px 0 0;}">
    <el-pagination
      @size-change="handleSizeChange"
      @current-change="handleCurrentChange"
      :current-page="pageNum"
      :page-sizes="[5, 10, 15, 20]"
      :page-size="pageSize"
      layout="total, sizes, prev, pager, next, jumper"
      :total="total">
    </el-pagination>
</div>

<script>
    async getTableData(){
        const GetCustomerListApiRes = await GetCustomerListApi({
            pageNum:this.pageNum,
            pageSize:this.pageSize
        });
        console.log(GetCustomerListApiRes);
        this.tableData = GetCustomerListApiRes.rows;
        //获取表格总数据
        this.total=GetCustomerListApiRes.total;
    },
    handleCurrentChange(val){
        //当前页数改变时执行这个函数
        console.log("页数改变",val);
        this.pageNum = val;
        this.getTableData();
    },
    handleSizeChange(val){
        //每页条数改变时执行这个函数
        console.log("每页条数改变",val);
        this.pageSize = val;
        this.getTableData();
    }
</script>
```

## 9.27 导出文件与上传文件的处理

### 9.27.1 导出文件

以导出客户管理模块中的客户信息为例,在 api.js 文件中,定义导出客户信息表格方法,代码如下:

```
//导出客户信息表格
export const CustomerExportApi = (params,configs) => request.post("/prod-api/customer/export",params,configs);
```

这里在请求时要小心,后端没有直接返回 code 字段,而是文件的二进制字符,所以需要在响应拦截器中的错误响应判断前添加判断 code 是否存在的代码,代码如下:

```
if(data.code && data.code!==200){
    ...
    return false
}
```

安装保存文件的插件 file-saver,命令如下:

```
npm install file-saver --save
```

在组件中实现文件保存功能,代码如下:

```
<!-- 第 9 章 Vue.js 2.0 全家桶+Element 开发后台管理系统:导出文件与上传文件的处理 -->
import {saveAs} from 'file-saver';
import {CustomerExportApi} from "@/request/api"
export default {
    methods:{
        async exportExl(){
            let res = await CustomerExportApi({
                pageNum: 1,
                pageSize: 10
            },{
                //导出文件的请求需要额外传请求配置
                headers: { 'Content-Type': 'application/x-www-form-urlencoded'},
                responseType: 'blob'
            });
            saveAs(
                //Blob 对象表示一个不可变、原始数据的类文件对象
                //Blob 表示的不一定是 JavaScript 原生格式的数据
                //File 接口基于 Blob,继承了 Blob 的功能并对其进行扩展,使其支持用户系
                //统上的文件
                //返回一个新创建的 Blob 对象,其内容由参数中给定的数组串联组成
                new Blob([res]),
                //设置导出文件名称
```

```
                `客户档案_${new Date().getTime()}.xlsx`
            );
        }
    }
}
```

## 9.27.2 上传文件

在 API 中,实现上传组件的页面结构与页面功能,代码如下:

```
//审核流程定义
export const BpmnInfoApi = (params,configs) => request.post("/prod-api/business/bpmnInfo",params,configs);
```

组件中的代码如下:

```
<!-- 第 9 章 Vue.js 2.0 全家桶+Element 开发后台管理系统:导出文件与上传文件的处理 -->
<template>
    <div>
        <el-upload
        class="upload-demo"
        ref="upload"
        :file-list="fileList"
        :auto-upload="false"
        :on-change="handleChange">
        <el-button slot="trigger" size="small" type="primary">选取文件</el-button>
        <el-button style="margin-left: 10px;" size="small" type="success" @click="submitUpload">上传到服务器</el-button>
        <div slot="tip" class="el-upload__tip">只能上传 bpmn 文件,并且不超过 500KB</div>
        </el-upload>
    </div>
</template>

<script>
import {BpmnInfoApi} from "@/request/api"
export default {
    data() {
        return {
            fileList: []
        };
    },
    methods: {
        async submitUpload() {

            const file = this.fileList[0].raw
```

```
            console.log(file);
            const BpmnInfoApiRes = await BpmnInfoApi({
                file:file,
                bpmnLabel:"试试",
                bpmnType:"casual-leave",
                info:"试试就试试"
            },{
                headers: {
                    'Content-Type': 'multipart/form-data'
                }
            })
            if(!BpmnInfoApiRes)return;
            this.$message.success(BpmnInfoApiRes.msg)

        },
        handleChange(file, fileList) {

            this.fileList = fileList
        },
    }
}
</script>
```

常见的 Content-Type 类型：post 请求的消息主体放在 entity body 中，服务器端根据请求头中的 Content-Type 字段获取消息主体的编码方式，进而对数据进行解析。

### 1. application/x-www-form-urlencoded

最常见的 post 提交数据的方式是原生 Form 表单，如果不设置 enctype 属性，则默认以 application/x-www-form-urlencoded 方式提交数据。

首先，Content-Type 被指定为 application/x-www-form-urlencoded；其次，提交的表单数据会转换为键-值对并按照 key1=val1&key2=val2 的方式进行编码，key 和 val 都进行了 URL 转码。大部分服务器端语言对这种方式有很好的支持。

另外，当利用 AJAX 提交数据时，也可使用这种方式。例如 jQuery 和 Content-Type 的默认值都是 application/x-www-form-urlencoded;charset=utf-8。

### 2. multipart/form-data

另一个常见的 post 数据提交的方式是将 Form 表单的 enctype 设置为 multipart/form-data，它会将表单的数据处理为一条消息，以标签为单元，用分隔符（这就是 boundary 的作用）分开，类似上面 Content-Type 中的例子。

由于这种方式将数据分为很多部分，它既可以上传键-值对，也可以上传文件，甚至多个文件。当上传的字段是文件时，会有 Content-Type 来说明文件类型；Content-disposition 用来说明字段的一些信息。每部分都是以 -boundary 开始，紧接着是内容描述信息，然后是回车，最后是字段的具体内容（字段、文本或二进制等）。如果传输的是文件，则要包含文件名和文件类型信息。消息主体最后以 -boundary- 标示结束。

### 3. application/json

Content-Type: application/json 作为响应头比较常见。实际上，现在越来越多的人把它作为请求头，用来告诉服务器端消息主体是序列化后的 JSON 字符串，其中一个好处就是 JSON 格式支持比键-值对复杂得多的结构化数据。由于 JSON 规范的流行，除了低版本 IE 之外的各大浏览器都原生支持 JSON.stringify，服务器端语言也都有处理 JSON 的函数，使用起来没有困难。

谷歌的 AngularJS 中的 AJAX 功能，默认就是提交 JSON 字符串。

### 4. text/xml

XML 的作用不言而喻，用于传输和存储数据，它非常适合万维网传输，提供统一的方法来描述和交换独立于应用程序或供应商的结构化数据，在 JSON 出现之前是业界一大标准（当然现在也是），相比 JSON 的优缺点有兴趣的读者可以上网查找，因此，在 post 提交数据时，XML 类型也是不可缺少的一种，虽然一般场景上使用 JSON 可能更轻巧、更灵活。

### 5. binary (application/octet-stream)

在 Chrome 浏览器的 Postman 工具中，还可以看到 binary 这一类型，指的就是一些二进制文件类型。如 application/pdf，指定了特定二进制文件的 MIME 类型。就像对于 text 文件类型，若没有特定的子类型（subtype），就使用 text/plain。类似地，二进制文件没有特定或已知的 subtype，即使用 application/octet-stream，这是应用程序文件的默认值，一般很少直接使用。

对于 application/octet-stream，只能提交二进制，而且只能提交一个二进制，如果提交文件，只能提交一个文件，后台接收参数只能有一个，而且只能是流（或者字节数组）。

很多 Web 服务器使用默认的 application/octet-stream 来发送未知类型。出于安全原因，对于这些资源浏览器不允许设置一些自定义默认操作，导致用户必须存储到本地以使用。一般来讲，设置正确的 MIME 类型很重要。

# 第 10 章 Git 介 绍

目前企业多用 Git 做代码版本控制,熟练使用 Git 已经是 IT 开发人员的一项必备技能。Git 是一个免费的开源的分布式版本控制系统,它可以快速高效地处理从小型到大型的项目。简单来说 Git 就是分布式版本控制工具(管理代码版本)。常见版本控制系统有 Git、SVN 等,如图 10-1 和图 10-2 所示。常用的 Git 代码托管平台有 Gitee(码云)、GitLab、GitHub 等。

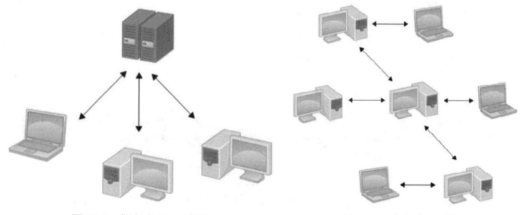

图 10-1 集中式 SVN 模型　　　　图 10-2 分布式 Git 模型

分布式与集中式模型的对比如下:
(1) Git 是分布式的,SVN 是集中式的。
(2) Git 每个历史版本存储完整的文件,而 SVN 则只存储文件的差异。
(3) Git 可离线完成大部分操作,SVN 则必须与中央服务器进行网络交互。
(4) Git 有着优雅的分支和合并功能。
(5) Git 有着更强的撤销修改和修改版本历史的能力。
(6) Git 速度更快,效率更高。

## 10.1 Git 的基本使用

**1. 下载及安装 Git**

Git 的下载网址为 https://git-scm.com/downloads，安装成功后进行检查。

方式一：通过快捷键 Win+R 打开运行，输入 cmd 打开控制台，在控制台输入 git --version 命令查看 Git 版本。

方式二：右击出现两个 Git 菜单。

**2. Git 全局配置**

按快捷键 Win+R 打开运行窗口，输入 cmd，然后按 Enter 键，打开控制台窗口，输入的命令如下：

```
<!-- 第 10 章 Git 介绍：Git 的基本使用 -->
#基本配置：user.name & user.email
git config --global user.name "Your Name"
git config --global user.email "email@example.com"
#这台机器上所有的 Git 仓库都会使用这个配置

#查看配置：
git config --list
```

**3. Gitee 远程创建仓库**

在 https://gitee.com/ 上进行注册并登录。单击"+"按钮新建仓库，如图 10-3 所示。

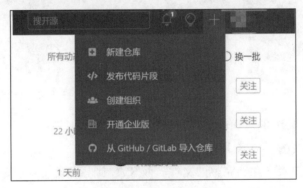

图 10-3　新建远程仓库

添加仓库信息，如图 10-4 所示。

**4. 在 Gitee 中配置 SSH 公钥**

注册（用邮箱注册）https://gitee.com/，并登录到 Gitee 后，将鼠标移至右上角头像下拉选项中选择设置，在左侧菜单栏中选择 SSH 公钥，在右侧添加公钥。怎么生成本机的公钥？可查看 https://gitee.com/help/articles/4181。

图 10-4　添加仓库信息

创建 SSH Key：在黑窗口中键入命令 ssh-keygen -t rsa -C youremail@example.com，按照提示按下三次 Enter 键，即可生成 SSH Key。在本地用户目录下，找到本地文件 id_rsa.pub 的内容，复制生成的 SSH Key，选择"设置"→"安全设置"→"单击 SSH 公钥"，把生成的 Public Key 添加到仓库中。

### 5. Git 日常操作

Git 日常操作包含的一些命令如下。

（1）克隆代码（把远程仓库拉取到本地）：git clone 仓库地址。
（2）查看仓库状态：git status。
（3）将工作区代码提交到暂存区：①git add 文件路径；②.（所有文件）；③–all。
（4）将暂存区代码提交到本地仓库（历史记录）区：git commit -m '本次提交的信息提示。
（5）将本地仓库（历史记录）区的提交记录提交到远程仓库：git push origin 分支名称。
（6）查看提交的历史记录：git log 或　git log --pretty=oneline。
（7）简洁查看所有分支的所有操作记录：git reflog。

**注意**　Git 提示的信息里面只要有 fatal 或者 error 中的一个都表示执行 Git 命令失败了。

### 6. 工作区、暂存区、主分支

本地的 Git 的工作区、暂存区和主分支结构图如图 10-5 所示。

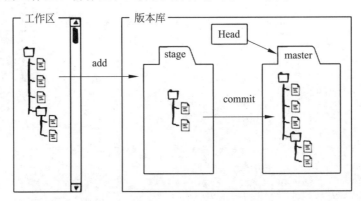

图 10-5　工作区、暂存区、主分支

工作区：计算机中能够看到的目录，例如 gitDemo 文件夹就是一个工作区；版本库：工作区有一个隐藏目录.git，这就是 Git 版本库。版本库中有一个暂存区 stage 自动生成一个 master 分支，指向一个 Head。

### 7. 分支

【**示例 10-1**】 分支使用场景：3月刚发布一个 v1.0 版本的项目，发布完之后张三开始继续开发 A 功能，当开发到 50% 时，发现 v1.0 版本项目出 Bug 了，需要立即修复。如果张三改的正好就是 v1.0 版本代码，则必须把代码删掉，先修复 Bug。

解决方案及命令如下：

```
查看分支：git branch
创建分支：git branch <name>
切换分支：git checkout <name>
创建&切换分支：git checkout -b <name>
删除分支：git branch -d <名称>
```

在 Gitee 中创建项目，例如创建了 demo0630 项目，命令如下：

```
//将仓库克隆到本地
git clone git@gitee.com:xxxx/xxxx.git
//检查当前分支
git branch
```

一般此时是 master 分支，但开发阶段一般用 dev 等其他分支，所以需要切换到 dev 分支，命令如下：

```
//创建并切换到 dev 分支
git checkout -b dev
```

注意：
（1）在仓库没有初始化前，如果需要检查分支，则一定要记住这句话，即未曾 commit 的仓库是无法检查分支的。
（2）在切换分支之前要保证工作树是干净的。
（3）实际上使用 git switch 命令也可以实现创建和切换分支功能，命令如下：

```
git switch -c dev      //创建并切换到新的 dev 分支
git switch dev         //切换到已有的 dev 分支
```

不同分支下，提交文件的方式相同，git add .、git commit -m " "、git push origin dev，此时会发现，git checkout master 切换回 master 分支后，dev 分支下修改的内容看不见，因为当前分支不存在这个修改动作。

### 8. 合并分支

假设当前项目已经完成，想要把 dev 分支合并到 master，则应先切换到 master 分支，再通过 merge 合并 dev 分支，命令如下：

```
Git checkout master
git merge dev
```

### 9. 合并时出现冲突问题

在 master 分支下和 dev 分支下同时对 index.js 文件进行修改，并提交到本地仓库。切换到 master 分支中合并 dev 分支，由于都针对同样的问题进行了修改，此时会出现冲突。

### 10. 版本回滚

使用 git switch -c dev1 命令创建新的 dev1 分支，稍微修改 dev1 中的 index.js 文件，并且提交到远程仓库。

提交完成后，如果不想用当前代码，想回滚到上一次的代码，则命令如下：

```
<!-- 第10章 Git 介绍：Git 的基本使用 -->
//查看当前项目提交过的所有版本（含所有分支操作）
git log
//如果只想简单地看一看版本号，可以使用以下命令
git log --pretty=oneline
```

结果如下：

```
<!-- 第10章 Git 介绍：Git 的基本使用 -->
b2ff1beb92bd3ac425dac2fa519d4b8191438be9 (HEAD -> dev1) '123456'
c9efd011b471765fc9fdd6eefaadf75b3b36153b (origin/dev1) '12345'
80f6cb77a9b13cd471b725b5ef66901150bf57bb '提交'
47d676dbf6d0687d76f059ee9ed044c4c378ed30 (origin/master, origin/HEAD) 'dev1 的首次提交'
d5c756afd04abfdbc9bb2299064eab34b30ede5f (origin/dev, dev) '修改了 index.js'
```

```
65f4c72f6c8e2c272d4e284c103e249c65ebff32 Initial commit
```

如果只想回滚到指定版本，则可通过 reset 实现版本回滚，命令如下：

```
git reset --hard c9efd
```

这里只需写 id 的前几个字母与数字，没必要全写，Git 会自动去检索，但此时如果再一次查看所有版本，则会发现：

```
<!-- 第 10 章 Git 介绍：Git 的基本使用 -->
c9efd011b471765fc9fdd6eefaadf75b3b36153b (origin/dev1) '12345'
80f6cb77a9b13cd471b725b5ef66901150bf57bb '提交'
47d676dbf6d0687d76f059ee9ed044c4c378ed30 (origin/master, origin/HEAD)
'dev1 的首次提交'
d5c756afd04abfdbc9bb2299064eab34b30ede5f (origin/dev, dev) '修改了 index.js'
65f4c72f6c8e2c272d4e284c103e249c65ebff32 Initial commit
```

最新写的那个丢失了，但如果此时又后悔了，则该怎么办？可以通过 reset 重新找回，命令如下：

```
//重新 reset 即可找回
git reset --hard b2ff1
```

现在，就可以退回最新版本了。

但这种方法有效的前提是当前这个控制台不曾关掉，如果已经关掉了，则没法知道版本号，这时该怎么办呢？命令如下：

```
//Git 提供了一个命令 git reflog，用来记录每次命令
git reflog
```

## 10.2 Git Flow 工作流模型

Git Flow 定义了一个项目发布的分支模型，为管理具有预定发布周期的大型项目提供了一个健壮的框架，如图 10-6 所示。此工作流模型中要求有以下分支类型。

（1）feature 分支：用于功能开发。
（2）develop 分支：用于聚合 feature 分支开发的功能。
（3）release 分支：用于测试发行版。
（4）master 分支：打上版本 Tag 长期稳定支持，任何一个 Tag 都可以稳定发布。
（5）hotfixes 分支：用于修复线上 Bug。

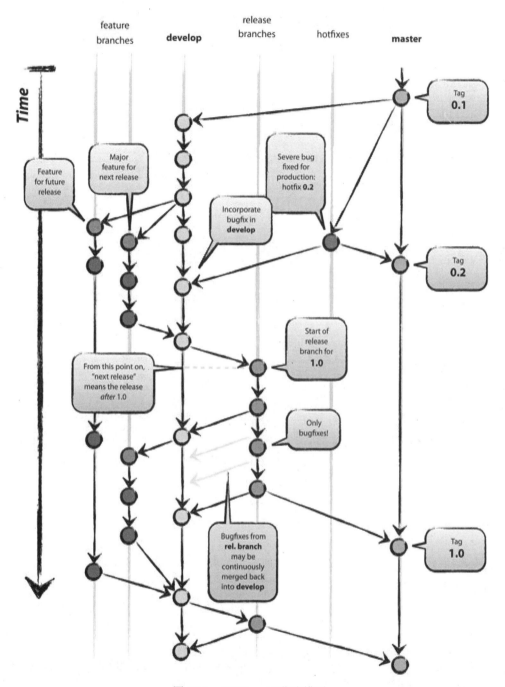

图 10-6 Git Flow 工作流模型

## 10.3　Git 拓展

**1. git fetch**

git fetch 命令用于从远程获取代码库，命令如下：

```
<!-- 第 10 章 Git 介绍：Git 拓展 -->
#获取远程仓库:git fetch 远程分支
git fetch origin master

#对比区别
git diff origin/master

#合并远程仓库
git merge origin/master
```

**2. git pull**

git pull 命令相当于执行 git fetch 命令以后，接着执行 git merge 命令。就是它把下载与合并这两个动作结合到一块儿了，命令如下：

```
<!-- 第 10 章 Git 介绍：Git 拓展 -->
git fetch origin
git merge origin/master
#获取远程仓库
git pull 远程
```

**3. git fetch 和 git pull 的区别**

在作用上，它们的功能是大致相同的，都起到了从远程仓库下拉代码的作用，但是两者也有区别，git pull 命令相当于先执行 git fetch 命令，然后执行 git merge 命令对代码进行合并。

可以随时执行 git fetch 命令来更新 refs/remotes/<remote>/下的远程 tracking 分支，但 fetch 操作不会更改 refs/heads 下的本地分支，也不会更改工作副本，这个操作是安全的，而 git pull 操作则会把远程版本的最新更改更新到本地分支，同时还更新其他远程跟踪分支。

**4. git rebase 变基**

rebase 可翻译为变基，它的作用和 merge 很相似，用于把一个分支的修改合并到当前分支上。

当执行 rebase 操作时，Git 会从两个分支的共同祖先开始提取待变基分支上的修改，然后将待变基分支指向基分支的最新提交，最后将刚才提取的修改应用到基分支的最新提交的后面，如图 10-7 所示。

准备工作如下。

（1）master: readme.md。

（2）创建开发分支 dev。

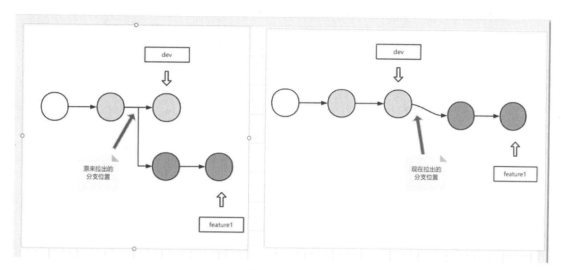

图 10-7　变基

（3）创建功能分支 feature1。

（4）切换到 dev：新增 index.html 文件后提交，此处模拟的是同事 B 在 feature2 中已经完成，并已合并到 dev。

（5）切换到 feature1，创建 goods.vue 文件后提交，创建 goods.js 文件后提交。

（6）在 feature1 中执行 git rebase dev 变基。

注意：在大部分情况下，rebase 的过程中会产生冲突，此时，就需要手动解决冲突问题，然后依次使用 git add、git rebase --continue 的方式来处理冲突问题，完成 rebase 的过程。如果不想要某次 rebase 的结果，则需要使用 git rebase --skip 来跳过这次 rebase 操作。

冲突后，先通过 git add <file-name>命令来标记冲突已解决，再使用 git rebase --continue 继续，如果中间遇到某个补丁不需要应用，则可以用命令 git rebase --skip 忽略。如果想回到 rebase 执行之前的状态，则可以执行 git rebase --abort 命令。

### 5. git rebase 和 git merge 的区别

不同于 git rebase 命令的是，git merge 命令在不是 fast-forward（快速合并）的情况下，会产生一条额外的合并记录。另外，在解决冲突时，用 merge 只需解决一次冲突，而用 rebase 时，则需要依次解决每次冲突后才可以提交。

git merge 命令如图 10-8 所示。

情况 1：在 feature1 中执行 git rebase dev 变基；情况 2：在 feature1 中执行 git merge dev，命令如下：

```
git log --graph
git log --oneline
```

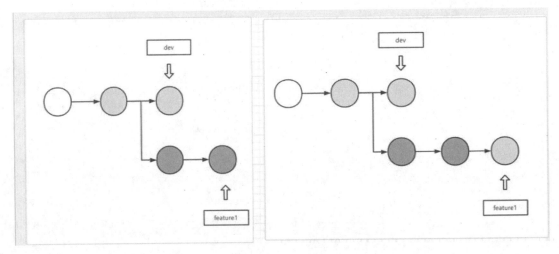

图 10-8　git merge

#### 6. git rebase 合并多次提交

在开发中，常会遇到在一个分支上产生了很多无效的提交，在这种情况下使用 rebase 的交互式模式可以把已经发生的多次提交压缩成一次提交，得到了一个干净的提交历史，如图 10-9 所示。

```
$ git log --pretty=oneline
1e9f3a1f0ca9634417e883a50526f3c5acc25caf (HEAD -> master) 增加footer
9a9286e5d384a63767727ed0f4b662e57bfa3b5f 增加sections
27da58b15e84f09c672fc3d97d45a304c54b48a4 初始化html
ec3f5c7b1b0c684212c4adb1d30f88b73480a2fe (origin/master, origin/HEAD) Initial co
mmit
```

图 10-9　git rebase 合并多次提交（1）

【示例 10-2】将 HTML 相关的提交合并成一个。

命令如下：

```
git rebase -i <base-commit>
git rebase -i HEAD~2
```

参数 base-commit 就是指明操作的基点提交对象，基于这个基点进行 rebase 操作，对于上述提交历史的例子，要把最后的一个提交对象（9d4cf2f）之前的提交压缩成一次提交，命令如下：

```
git rebase -i ec3f5c
```

此时会进入一个 vim 的交互式页面，编辑器列出的信息如图 10-10 所示。

想要合并这一堆更改，要使用 Squash 策略进行合并，即把当前的 commit 和它的上一个 commit 内容进行合并，大概可以表示为下面这样，在交互模式的 rebase 下，至少保留一个 pick，否则命令会执行失败。

```
pick 27da58b 初始化html
pick 9a9286e 增加sections
pick 1e9f3a1 增加footer

# Rebase ec3f5c7..1e9f3a1 onto ec3f5c7 (3 commands)
#
# Commands:
# p, pick <commit> = use commit
# r, reword <commit> = use commit, but edit the commit message
# e, edit <commit> = use commit, but stop for amending
# s, squash <commit> = use commit, but meld into previous commit
# f, fixup <commit> = like "squash", but discard this commit's log message
# x, exec <command> = run command (the rest of the line) using shell
# b, break = stop here (continue rebase later with 'git rebase --continue')
# d, drop <commit> = remove commit
# l, label <label> = label current HEAD with a name
# t, reset <label> = reset HEAD to a label
# m, merge [-C <commit> | -c <commit>] <label> [# <oneline>]
# .       create a merge commit using the original merge commit's
# .       message (or the oneline, if no original merge commit was
# .       specified). Use -c <commit> to reword the commit message.
#
# These lines can be re-ordered; they are executed from top to bottom.
<t101701C/.git/rebase-merge/git-rebase-todo [unix] (20:14 17/10/2022)1,1 顶端
<test101701C/.git/rebase-merge/git-rebase-todo" [unix] 28L, 1173C
```

图 10-10 git rebase 合并多次提交（2）

（1）需要保留的是第 1 个，保持前面的命令 pick 不动。
（2）将第 2 行和第 3 行的命令 pick 改成 squash。
（3）输入 wq，保存并退出。

pick：保留该 commit（缩写为 p）。

reword：保留该 commit，但需要修改该 commit 的注释（缩写为 r）。

edit：保留该 commit，但要停下来修改该提交，不仅修改注释（缩写为 e）。

squash：将该 commit 和前一个 commit 合并（缩写为 s）。

fixup：将该 commit 和前一个 commit 合并，但不要保留该提交的注释信息（缩写为 f）。

exec：执行 Shell 命令（缩写为 x）。

drop：要丢弃该 commit（缩写为 d）。

label：用名称标记当前 HEAD（缩写为 l）。

reset：将 HEAD 重置为标签（缩写为 t）。

merge：创建一个合并分支并使用原版分支的 commit 的注释（缩写为 m）。

输入 wq，保存并退出后，又出现一个编辑框，里面是需要保留的 commit 信息，如图 10-11 所示。

把不需要的 commit 信息直接删掉，保留第 1 个的 commit 信息（也可以修改一下），再次输入 wq 保存并退出，如图 10-12 所示。

接下来查看日志，是不是清爽了很多？rebase 操作可以让提交历史变得更加清晰，如图 10-13 所示。

### 7. git stash 暂存文件

git stash 命令可以将当前的工作状态保存到 Git 栈，在需要时再恢复。

```
# This is a combination of 3 commits.
# This is the 1st commit message:

初始化html

# This is the commit message #2:

增加sections

# This is the commit message #3:

增加footer

# Please enter the commit message for your changes. Lines starting
# with '#' will be ignored, and an empty message aborts the commit.
#
# Date:      Mon Oct 17 19:29:34 2022 +0800
#
# interactive rebase in progress; onto ec3f5c7
# Last commands done (3 commands done):
#     squash 9a9286e 增加sections
#     squash 1e9f3a1 增加footer
# No commands remaining.
<码/ceshi3/test101701C/.git/COMMIT_EDITMSG" [unix] (20:15 17/10/2022)1,1 顶端
<04-代码/ceshi3/test101701C/.git/COMMIT_EDITMSG" [unix] 28L, 664C
```

图 10-11　git rebase 合并多次提交（3）

```
# This is a combination of 3 commits.
# This is the 1st commit message:
初始化html,sections,footer
# This is the commit message #2:
# This is the commit message #3:
# Please enter the commit message for your changes. Lines starting
# with '#' will be ignored, and an empty message aborts the commit.
#
# Date:      Mon Oct 17 19:29:34 2022 +0800
#
```

图 10-12　git rebase 合并多次提交（4）

```
$ git log --pretty=oneline
fdf05e5b456e1358d80676b77913a8550c25132b (HEAD -> master) 初始化html,sections
,footer
ec3f5c7b1b0c684212c4adb1d30f88b73480a2fe (origin/master, origin/HEAD) Initial
commit
```

图 10-13　git rebase 合并多次提交（5）

　　会有这么一个场景，现在正在你的 feature 分支上开发新功能。这时，生产环境上刚好出现了一个 Bug 需要紧急修复，但是这部分代码还没开发完，不想提交，怎么办？这时可以用 git stash 命令先把工作区已经修改的文件暂存起来，然后切换到 hotfix 分支上进行 Bug 的修复，修复完成后，切换回 feature 分支，从堆栈中恢复刚刚保存的内容，命令如下：

```
<!-- 第10章 Git 介绍：Git 拓展 -->
git stash          //把本地的改动暂存起来，保存当前的工作区与暂存区的状态，把当前的修改保存到
                   //Git 栈，等以后需要时再恢复，git stash 命令可以多次使用，每次使用都会新加
                   //一个 stash@{num}，num 是编号
git stash save "message"   //执行存储时，添加备注，方便查找。作用等同于 git stash,
                           //区别是可以加一些注释，执行存储时，添加注释，方便查找
git stash pop      //应用最近一次暂存的修改，并删除暂存的记录
```

```
git stash list   //查看 stash 有哪些存储
git stash apply //应用某个存储，但不会把存储从存储列表中删除，默认使用第 1 个存储，
                //即 stash@{0}，如果要使用其他的，则使用 git stash apply
                //stash@{$num}
git stash clear //删除所有缓存的 stash
```

## 第 11 章

# 项目二：大型 PC 商城

第 2 个 Vue.js 项目是"叩丁严选"大型 PC 商城，是一个由 Vue CLI 搭建的 PC 端 SPA 商城，该商城主要涉及登录注册、商品列表、商品详情、个人中心、购物车及商品检索等主体功能。该项目主要用于平台用户参与积分兑换商品，是一个大型的 PC 端商城项目。

## 11.1 项目准备

### 1. 创建项目
通过以下命令在自己的目下创建项目：

```
> vue create v2-shop
```

### 2. 清空项目非必要文件
（1）views 下面的文件只保留 Home.vue，其余删除。删除 components/HelloWorld.vue，并且 Home.vue 文件中不再引入 HelloWorld 组件。
（2）删除 src/assets 下的图片，换成 img 文件夹。
（3）将 router/index.js 文件中的 about 路由注释掉。
（4）删除 App.vue 文件中的 less。

### 3. 样式初始化
在 App.vue 组件中书写结构样式，代码如下：

```
<!-- 第11章大型PC商城：项目准备 -->
<div class="header"></div>

</script>
<style lang="less" scoped>
.header{
  height: 50px;
  background-color: #333;
}
</style>
```

发现页面会有自带的间距，这是浏览器本身的默认样式，所以需要进行样式初始化，清除浏览器默认样式。

安装初始化样式库 reset-css，命令如下：

```
npm i reset-css 或者  yarn add reset-css
```

安装成功后在 main.js 文件中引入即可，代码如下：

```
import "reset-css"
```

4. 网站结构布置

在 App.vue 文件中设置好头部、导航和尾部组件，代码如下：

```
<!-- 第11章大型PC商城：项目准备 -->
<template>
  <div id="app">
      <Tabbar></Tabbar>
      <Header></Header>
      <router-view/>
      <Footer></Footer>
  </div>
</template>
<script>
import Header from '@/components/Header'
import Tabbar from '@/components/Tabbar'
import Footer from '@/components/Footer'

export default {
   components:{
       Header,Tabbar,Footer
   },

}
</script>
```

在 @/components 目录下新建 Header、Tabbar、Footer 这 3 个组件即可。

## 11.2 网站数据请求模块

接口文档的网址为 http://www.docway.net/project/1h9xcTeAZzV/share/1iUU09vKhMm。接口的网址为 http://kumanxuan1.f3322.net:8881/cms。

1. 发起请求

作为一个网站前端，数据请求模块少不了。安装 axios 模块，命令如下：

```
npm i axios
```

尝试在 app.vue 文件中进行数据请求，代码如下：

```html
<!-- 第11章大型 PC 商城：网站数据请求模块 -->
```
```js
import axios from "axios"//引入 axios 模块
export default {
    ...
  created(){
    //网络请求
    axios.get("http://192.168.113.249:8081/cms/products/recommend")
    .then(res=>{
        //请求成功
        console.log(res.data);
    })
    .catch(err=>{
        //请求失败
        console.log(err);
    })
  },
}
```

### 2. 代理配置

对 vue.config.js 文件进行配置，代码如下：

```html
<!-- 第11章大型 PC 商城：网站数据请求模块 -->
```
```js
module.exports = {
    devServer: {//配置代理
        port: 8080,
        proxy: {
            '/api': {//请求前缀
                target: "http://192.168.113.249:8081/cms",//目标地址
                pathRewrite: {
                    '^/api': ''
                }
            }
        }
    }
}
```

由于配置文件修改了，所以这里一定要记得重新通过 yarn serve 命令运行项目。

### 3. API 与 Request 封装

在 src 下新建 request 目录，在 request 目录下新建 request.js 文件。

在 request.js 文件中，创建 axios 并对 axios 进行请求配置，代码如下：

```html
<!-- 第11章大型 PC 商城：网站数据请求模块 -->
```
```js
import axios from "axios"//引入 axios 模块
//创建 axois 实例
const instance = axios.create({
    baseURL:"http://192.168.113.249:8081/cms",
    timeout:5000
```

```
      })
      //请求配置
      instance.interceptors.request.use(config=>{
          console.log("每次发起请求前都会先执行这里的代码");
          console.log(config);   //本次请求的配置信息
          return config
      },err=>{
          return Promise.reject(err)
      })
      //响应配置
      instance.interceptors.response.use(res=>{
          console.log("每次接收到响应都会先执行这里的代码,再去执行成功的那个回调函数then");
          return res
      },err=>{
          return Promise.reject(err)
      })

      export default instance
```

为了更好地管理这些接口,先把所有请求都抽取出来,然后存放在一个 api.js 文件中。

在 request 目录下新建 api.js 文件,在 api.js 文件中引入请求模块,定义请求精品推荐模块数据,代码如下:

```
import request from './request'//引入 request

//请求精品推荐数据
export const JingpinAPI = () => request.get('/products/recommend')
```

### 4. 发起请求

在 App.vue 文件中,引入精品推荐模块数据请求方法,获取精品推荐模块数据,代码如下:

```
<!-- 第 11 章大型 PC 商城:网站数据请求模块 -->
import {JingpinAPI} from "@/request/api"

created(){
    JingpinAPI()
        .then(res=>{
            if(res.errno == 0){
                console.log(res.data)//成功获得所有首页数据
            }
        })
}
```

## 11.3 头部组件

### 11.3.1 版心样式

在 assets 下新建 css 目录,新建 public.less 文件,代码如下:

```less
.wrap{
    width: 1200px;
    margin: 0 auto;
}
```

在 main.js 文件中引入 public.less 文件,代码如下:

```js
import "@/assets/css/public.less"
```

### 11.3.2 头部组件布局

头部组件布局,代码如下:

```html
<!-- 第11章大型PC商城:头部组件 -->
<template>
    <div class="header">
        <div class="wrap header-wrap">
            <div class="l">
                欢迎来到叩丁严选
            </div>
            <div class="r">
                <ul>
                    <li class="avatar">
                        <img src="../assets/img/userImg.f8bbec5e.png" width="26" alt="">
                        用户名: --
                    </li>
                    <li>
                        我的鸡腿: --
                    </li>
                    <li>获取鸡腿</li>
                    <li>叩丁狼官网</li>
                    <li class="login-btn">登录</li>
                </ul>
            </div>
        </div>
    </div>
</template>

<script>
```

```
export default {
    data () {
        return {


        }
    }
}
</script>

<style lang = "less" scoped>
    .header{
        height: 40px;
        background: #242B39;

        .header-wrap{
            color:#FFFEFE;
            height: 40px;
            font-size: 14px;
    /*      background-color: #fcf; */
            display: flex;
            justify-content: space-between;
            align-items: center;
            .r ul{
                display: flex;
                align-items: center;
                li{
                    margin-right: 20px;
                    cursor: pointer;
                }
                .avatar{
                    display: flex;
                    align-items: center;

                    img{
                        border-radius: 50%;
                        margin-right: 5px;
                    }
                }

                .login-btn{
                    width: 124px;
                    height: 40px;
                    text-align: center;
                    line-height: 40px;
                    background: #0A328E;
                }
```

```
                    }
                }
            }
</style>
```

## 11.4 导航组件

### 11.4.1 基本布局

导航基本布局效果如图 11-1 所示。

图 11-1 导航基本布局

在 Nav.vue 组件中渲染导航组件，代码如下：

```
<!-- 第 11 章大型 PC 商城：导航组件 -->
<template>
    <div class="nav">
        <div class="wrap nav-wrap">
            <div class="l">
            <h1>
                <img src="../assets/img/indexLogo.6f8ac4f0.png" alt="">
            </h1>
            </div>
            <div class="c">
                <ul>
                    <li>首页</li>
                    <li>全部商品</li>
                    <li>个人中心</li>
                    <li>我的订单</li>
                    <li>专属福利</li>
                </ul>
            </div>
            <div class="r">
                <input type="text">
                <span><img src="../assets/img/search.png" alt=""></span>
            </div>
        </div>
    </div>
</template>

<script>
```

```
export default {
    data () {
        return {


        }
    }
}
</script>

<style lang = "less" scoped>
    .nav-wrap{
        height: 118px;
        background-color: #fcf;
        display: flex;
        justify-content: space-between;
        align-items: center;
        .c ul{
            width: 500px;
            display: flex;
            justify-content: space-between;
        }
    }
</style>
```

## 11.4.2 搜索框布局

搜索框基本布局效果如图 11-2 所示。

图 11-2 搜索框基本布局

```
<!-- 第 11 章大型 PC 商城：导航组件 -->
.r{
    display: flex;
    input{
        width: 214px;
        height: 40px;
        border: 1px solid #dedede;
        border-radius: 20px 0 0 20px;
        float: left;
        box-sizing: border-box;
        padding-left: 19px;
        outline-style: none;
```

```
    }
    .search-btn{
        width: 50px;
        height: 40px;
        background: #0A328E;
        border-radius: 0px 20px 20px 0px;
        text-align: center;
        line-height: 44px;
    }
}
```

### 11.4.3 路由配置及导航项当前样式

在 router/index.js 文件中配置重定向及几个导航路由，代码如下：

```
<!-- 第11章大型PC商城：导航组件 -->
const routes = [
  {
    path: '/',
    redirect: '/home'
  },
  {
    path: '/home',
    name: 'Home',
    component: Home
  },
  {
    path: '/goods',
    name: 'Goods',
    component: () => import(/* webpackChunkName: "goods" */ '../views/Goods.vue')
  },
  {
    path: '/user',
    name: 'User',
    component: () => import(/* webpackChunkName: "user" */ '../views/User.vue')
  },
  {
    path: '/order',
    name: 'Order',
    component: () => import(/* webpackChunkName: "dingdan" */ '../views/order.vue')
  },
  {
    path: '/free',
    name: 'Free',
```

```
      component: () => import(/* webpackChunkName: "free" */
'../views/Free.vue')
  }
]
```

在 View 目录中新建对应组件。处理导航项当前样式,在 Nav.vue 文件中,添加当前路由路径判断,代码如下:

```
<!-- 第 11 章大型 PC 商城:导航组件 -->
<div class="c">
    <ul>
        <li :class="$route.path==='/home'?'active':''">首页</li>
        <li :class="$route.path==='/goods'?'active':''">全部商品</li>
        <li :class="$route.path==='/user'?'active':''">个人中心</li>
        <li :class="$route.path==='/order'?'active':''">我的订单</li>
        <li :class="$route.path==='/free'?'active':''">专属福利</li>
    </ul>
</div>
<style>
...
    .c ul{
        width: 500px;
        display: flex;
        justify-content: space-between;
        color:#242B39;
        font-size: 16px;
        font-family: SourceHanSansSC-Medium;
        font-weight: 500;
        .active{
            color:#0A328E;
        }
    }
</style>
```

设置单击跳转路由,代码如下:

```
<!-- 第 11 章大型 PC 商城:导航组件 -->
<ul>
    <li @click="$router.push('/home')" :class="$route.path==='/home'?'active':''">首页</li>
    <li @click="$router.push('/goods')" :class="$route.path==='/goods'?'active':''">全部商品</li>
    <li @click="$router.push('/user')" :class="$route.path==='/user'?'active':''">个人中心</li>
    <li @click="$router.push('/order')" :class="$route.path==='/order'?'active':''">我的订单</li>
    <li @click="$router.push('/free')" :class="$route.path==='/free'?'active':''">专属福利</li>
</ul>
```

## 11.5 登录模块布局

### 11.5.1 模态窗口的书写

单击"登录"按钮,弹出模态窗口,如图11-3所示。

图11-3 模态窗口布局

在components目录下新建一个Login.vue组件,代码如下:

```
<!-- 第 11 章大型 PC 商城:登录模块布局 -->
<template>
    <div class="modal">
        <div class="mask"></div>
        <div class="login-box">

        </div>
    </div>
</template>

<script>
export default {}
</script>

<style lang = "less" scoped>
    .modal{
        position: fixed;
        width: 100%;
```

```
        height: 100%;
        left: 0;
        top: 0;
        /* bottom: 0;
        right: 0; */
        .mask{
            width: 100%;
            height: 100%;
            background-color: rgba(0,0,0,.5);
        }
        .login-box{
            position: absolute;
            left: 0;
            top: 0;
            bottom: 0;
            right: 0;
            margin:auto;
            width: 555px;
            height: 423px;
            background: url("../assets/img/login.png");
        }
    }
</style>
```

然后在 App.vue 组件中引入注册和使用，代码如下：

```
<!-- 第 11 章大型 PC 商城：登录模块布局 -->
<div id="app">
    <Header></Header>
    <Nav></Nav>
    <router-view/>
    <Footer></Footer>
    <Login></Login>
</div>
```

## 11.5.2 设置单击展示模态窗口

因为项目中可能会在各个组件中触发这个模态窗口进行展示，所以控制模态框展示的变量可以放在 Vuex 中。

在 store 中新建 showModal 目录，并在其中新建 index.js 文件，代码如下：

```
<!-- 第 11 章大型 PC 商城：登录模块布局 -->
export default{
    namespaced:true,
    state: {
        isShowLoginModal:false    //用来表示是否展示登录模态窗口
    },
```

```
    mutations: {
        //修改是否展示的值
        chanIsShowLoginModal(state,payload){
            state.isShowLoginModal = payload
        }
    },
    actions: {
    },
}
```

在 store.js 文件中引入 showModal，代码如下：

```
<!-- 第 11 章大型 PC 商城：登录模块布局 -->
import Vue from 'vue'
import Vuex from 'vuex'
import showModal from "./showModal"
Vue.use(Vuex)

export default new Vuex.Store({

  modules: {
    showModal
  }
})
```

在 Login.vue 组件中，添加是否显示登录模块功能，代码如下：

```
<!-- 第 11 章大型 PC 商城：登录模块布局 -->
<template>
    <div class="modal" v-show="isShowLoginModal">
        <div class="mask"></div>
        <div class="login-box">

        </div>
    </div>
</template>

<script>
import {mapState} from "vuex"
export default {
    computed:{
        ...mapState({
            isShowLoginModal:state=>state.showModal.isShowLoginModal
        })
    }
}
</script>
```

在 Header.vue 组件中，添加登录面板切换功能，代码如下：

```html
<!-- 第 11 章大型 PC 商城：登录模块布局 -->
<template>
    <div class="header">
        <div class="wrap header-wrap">
            ...
            <div class="r">
                <ul>
                    ...
                    <li class="login-btn" @click="chanIsShowLoginModal(true)">登录</li>
                </ul>
            </div>
        </div>
    </div>
</template>

<script>
import {mapMutations} from "vuex"
export default {
    data () {},
    methods:{
        ...mapMutations({
            chanIsShowLoginModal:"showModal/chanIsShowLoginModal"
        })
    }
}
</script>
```

### 11.5.3 单击关闭模态窗口

在 Login.vue 文件中，控制登录面板的关闭，代码如下：

```html
<!-- 第 11 章大型 PC 商城：登录模块布局 -->
<template>
    <div class="modal" v-show="isShowLoginModal">
        <!--单击遮罩层也可以关闭模态窗口-->
        <div class="mask"  @click="chanIsShowLoginModal(false)"></div>
        <div class="login-box">
            <div class="close" @click="chanIsShowLoginModal(false)"></div>
        </div>
    </div>
</template>
<script>
import {mapState,mapMutations} from "vuex"
export default {
```

```
        ...
        methods:{
            ...mapMutations({
                chanIsShowLoginModal:"showModal/chanIsShowLoginModal"
            })
        }
    }
</script>
<style lang = "less" scoped>
    .modal{
        ...
        .login-box{
            ...
            .close{
                width: 22px;
                height: 22px;
                background: url("../assets/img/close.png");
                position: absolute;
                right: 60px;
                top: 16px;
                cursor: pointer;
            }
        }
    }
</style>
```

## 11.5.4 单击标题栏的切换效果

在 Login.vue 文件中，添加标题栏切换功能，代码如下：

```
<!-- 第 11 章大型 PC 商城：登录模块布局 -->
<ul class="title">
    <li @click="isShowForm=true" :class="{active:isShowForm}">手机号码登录</li>
    <li style="margin:0 10px">|</li>
    <li @click="isShowForm=false" :class="{active:!isShowForm}">微信扫码登录</li>
</ul>
<div class="body">
    <div class="form" v-show="isShowForm">
        表单
    </div>
    <div class="qrcode" v-show="!isShowForm">
        二维码
    </div>
</div>
<script>
```

```
export default {
    data () {
        return {
            isShowForm:true
        }
    }
    ...
}
</script>
<style lang = "less" scoped>
    .modal{
        ...
        .login-box{
            ...
            .title{
                display: flex;
                justify-content: center;
                width: 100%;
                padding-top: 50px;
                font-size:20px;
                color:#999;
                .active{
                    color:#333;
                }
            }
            .body{
                width: 355px;
                margin:20px auto 0;
                height:200px;

            }
        }
    }
</style>
```

## 11.5.5 表单基本布局

在 Login.vue 文件中，实现表单基本布局，代码如下：

```
<!-- 第 11 章大型 PC 商城：登录模块布局 -->
<div class="body">
    <div class="form" v-show="isShowForm">
        <div class="mb20 row">
            <input type="text" class="ipt" placeholder="请输入手机号">
        </div>
        <div class="mb20 row">
            <input type="text" class="ipt" placeholder="请输入短信验证码">
```

```html
                <div class="btn checkcode-btn">获取验证码</div>
            </div>
            <div class="mb20 btn">
                登录
            </div>
        </div>
        <div class="qrcode" v-show="!isShowForm">
            二维码
        </div>
</div>
<style>
    ...
            .title{
                display: flex;
                justify-content: center;
                width: 100%;
                padding-top: 50px;
                font-size:20px;
                color:#999;
                .active{
                    color:#333;
                }
            }
            .body{
                width: 355px;
                margin:20px auto 0;
                height:200px;

                .form{
                    .row{
                        flex:1;
                        display: flex;

                    }
                    .ipt{
                        box-shadow: 0;
                        flex:1;
                        height: 48px;
                        border: 1px solid #e4e7eb;
                    }
                    .checkcode-btn{
                        width: 100px;
                        margin-left: 10px;
                    }
                    .btn{
                        background: #0a328e;
                        color: #fff;
```

```
                    text-align: center;
                    height: 50px;
                    line-height: 50px;
                    cursor: pointer;
                }
            }
        }
</style>
```

## 11.6 拼图验证滑块

插件参考的网址为 https://gitee.com/monoplasty/vue-monoplasty-slide-verify。

### 1. 安装插件
安装拼图验证滑块，命令如下：

```
npm install --save vue-monoplasty-slide-verify
//或者
yarn add vue-monoplasty-slide-verify
```

### 2. 引入拼图验证滑块
在 main.js 入口文件中引入拼图验证滑块，代码如下：

```
import SlideVerify from 'vue-monoplasty-slide-verify' //拼图验证码

Vue.use(SlideVerify)
```

### 3. 在组件中使用插件
在组件中使用拼图验证滑块组件，代码如下：

```
<!-- 第 11 章大型 PC 商城：拼图验证滑块 -->
<template>
        <slide-verify :l="42" :r="20" :w="362" :h="140" @success="onSuccess"
@fail="onFail" @refresh="onRefresh" :style="{ width: '100%' }" class="slide-box"
ref="slideBlock" :slider-text="msg"></slide-verify>
</template>

<script>
export default {
  data() {
    return {
      msg: "向右滑动"
    };
  },
  methods: {
    //拼图成功
    onSuccess(times) {
```

```
        let ms = (times / 1000).toFixed(1);
        this.msg = "login success, 耗时 " + ms + "s";
      },
      //拼图失败
      onFail() {
        this.onRefresh(); //重新刷新拼图
      },
      //拼图刷新
      onRefresh() {
        this.msg = "再试一次";
      },
    },
  };
</script>

<style lang="less" scoped>
/deep/.slide-box {
    width: 100%;
    position: relative;
    box-sizing: border-box;
    canvas {
        position: absolute;
        left: 0;
        top: -120px;
        display: none;
        width: 100%;
        box-sizing: border-box;
    }
    .slide-verify-block{
        width: 85px;
        height: 136px;
    }
    .slide-verify-refresh-icon {
        top: -120px;
        display: none;
    }
    &:hover {
        canvas {
            display: block;
        }
        .slide-verify-refresh-icon {
            display: block;
        }
    }
}
</style>
```

### 4. 单击"登录"按钮，判断是否有拼图滑块验证

登录之前，需要验证用户是否拼图验证过，只有拼图验证过才可以登录。以 msg 文字内容来判断是否有拼图滑块验证，代码如下：

```
<!-- 第 11 章大型 PC 商城：拼图验证滑块 -->
<div class="mb20 btn" @click="submitFn">
    登录
</div>
...
<script>
    ...
    methods:{
        ...
        //单击"登录"按钮
        submitFn() {
            //以msg文字内容来判断是否有拼图滑块验证
            if (this.msg == "再试一次" || this.msg == "向右滑动") {
               alert("请滑动拼图");
               return
            }
            alert("拼图滑块验证通过，可以执行登录了")
        },
    }
</script>
```

## 11.7　单击"获取验证码"按钮的逻辑

### 11.7.1　逻辑分析

可以正常获取验证码的前提是手机号格式正确，所以单击"获取验证码"按钮的逻辑如下：

（1）如果校验手机号格式不正确，则返回。
（2）滑块拼图验证不通过，则返回。
（3）验证成功后，发起请求，如果获取验证码成功，则进行倒计时。
结合运营商之后的手机号码的正则如下：

```
/^(13[0-9]|14[01456879]|15[0-35-9]|16[2567]|17[0-8]|18[0-9]|19[0-35-9])\d{8}$/
```

### 11.7.2　判断手机号格式

给"获取验证码"按钮添加获取验证码事件，实现手机号验证，代码如下：

```html
<!-- 第11章大型PC商城：单击"获取验证码"按钮的逻辑 -->
<div class="btn checkcode-btn" @click="getCode">获取验证码</div>
...
<script>
    getCode(){
        //验证手机号是否正确
        if(!/^(13[0-9]|14[01456879]|15[0-35-9]|16[2567]|17[0-8]|18[0-9]|19[0-35-9])\d{8}$/.test(this.phoneNum)){
            alert("请输入正确的手机号");
            this.$refs.phone.focus();
            return
        }
        alert("手机号格式正确");

        //进行滑块验证

        //验证成功后，发起请求，如果获取验证码成功，则进行倒计时，并展示秒数


    },
</script>
```

### 11.7.3 倒计时及其展示

当单击"获取验证码"按钮后，实现倒计时效果，代码如下：

```html
<!-- 第11章大型PC商城：单击"获取验证码"按钮的逻辑 -->
<div class="btn checkcode-btn" @click="getCode">
    <span v-show="!isShowCount">获取验证码</span>
    <span v-show="isShowCount">{{count}} s</span>
</div>
<script>
    methods:{
        countdown(){
            //计时的方法
            //倒计时，实际上就是每隔1s count 减去 1

            //每次单击先让 count 为 60
            this.count=60;
            let timer = null;
            timer = setInterval(()=>{
                this.count--
                if(this.count===0){
                    //清除定时器
                    clearInterval(timer)
```

```
                }
            },1000);
        },
        getCode(){
            //验证手机号是否正确
            /* if(!/^(13[0-9]|14[01456879]|15[0-35-9]|16[2567]|17[0-8]|
18[0-9]|19[0-35-9])\d{8}$/.test(this.phoneNum)){
                alert("请输入正确的手机号");
                this.$refs.phone.focus();
                return
            } */
            //进行滑块验证
            if (this.msg == "再试一次" || this.msg == "向右滑动") {
                alert("先进行滑块验证");
                return
            }
            //验证成功后，发起请求，如果获取验证码成功，则进行倒计时，并展示秒数
            //这里先展示秒数
            this.countdown();
            this.isShowCount=true;

        },
    }
</script>
```

## 11.7.4 连续单击倒计时 Bug

此时如果连续单击"获取验证码"按钮进行倒计时，则会有 Bug，数字越跳越快，主要是重复开启倒计时造成的。

其实只需把事件传给"获取验证码"所在的 span，就可以解决此问题，代码如下：

```
<!-- 第 11 章大型 PC 商城：单击"获取验证码"按钮的逻辑 -->
<div class="btn checkcode-btn">
    <span v-show="!isShowCount" @click="getCode">获取验证码</span>
    <span v-show="isShowCount">{{count}} s</span>
</div>
<script>
data () {
    return {
        //是否展示表单的布尔值
        isShowForm:true,
        //拼图滑块的文字
        msg: "向右滑动",
        //用户手机号
        phoneNum:"",
```

```
            //最大的计时时间
            countMax:60,
            //倒计时时间，每秒变化的那个数字
            count:0,
            //是否展示秒数所在盒子
            isShowCount:false
        }
    },
    ...
    countdown(){
        //计时的方法
        //倒计时，实际上就是每隔1s count 减去 1
        let timer = null;
        this.count = this.countMax;
        timer = setInterval(()=>{
            this.count--
            if(this.count===0){
                //清除定时器
                clearInterval(timer);
            }
        },1000);
    }
</script>
```

### 11.7.5 抽取工具函数

事实上，验证手机号本身是一个工具函数，和本身逻辑有关，但是具体代码不需要在组件中体现。真实项目场景往往是放在工具函数的文件中。

在 src 目录下新建 utils 文件夹，在里面新建 index.js 文件，添加手机号验证方法，代码如下：

```
<!-- 第 11 章大型 PC 商城：单击"获取验证码"按钮的逻辑 -->
export const validateTelephoneNumber = value =>{
    let reg = /^(13[0-9]|14[01456879]|15[0-35-9]|16[2567]|17[0-8]|18[0-9]|19[0-35-9])\d{8}$/
    return reg.test(value)
}
```

在 Login.vue 组件中，引入手机号验证方法，实现手机号验证，代码如下：

```
<!-- 第 11 章大型 PC 商城：单击"获取验证码"按钮的逻辑 -->
import {validateTelephoneNumber} from "@/utils"
...
getCode(){
```

```
    //验证手机号是否正确
    if(!validateTelephoneNumber(this.phoneNum)){
        alert("请输入正确的手机号");
        this.$refs.phone.focus();
        return
    }
    //进行滑块验证
        ...
    //验证成功后,发起请求,如果获取验证码成功,则进行倒计时,并展示秒数
        ...
},
```

## 11.7.6 发起获取验证码请求

在 api.js 文件中,添加发送短信验证请求方法,代码如下:

```
//发送短信验证码请求
export const SendSMSAPI = params => request.post("/sendSMS",params);
```

在 Login.vue 组件中,引入发送验证码请求方法,执行获取验证码请求。如果获取成功,则进行倒计时,代码如下:

```
<!-- 第 11 章大型 PC 商城:单击"获取验证码"按钮的逻辑 -->
//验证成功后,发起请求,如果获取验证码成功,则进行倒计时,并展示秒数
SendSMSAPI({
    phone:this.phoneNum.trim()
}).then(res=>{
    this.countdown();
    this.isShowCount=true;
    console.log(res);
})
```

返回的数据如图 11-4 所示。

```
▼{msg: "服务器出错,请您联系管理员", code: 500, message: "服务器出错,请您联系管理员", timestamp: 1632774621037}
    code: 500
    message: "服务器出错,请您联系管理员"
    msg: "服务器出错,请您联系管理员"
    timestamp: 1632774621037
```

图 11-4 验证码请求报错

接口上需要修改请求头 Content-Type 字段,并使用 qs.stringnify 进行格式转换,如图 11-5 所示。

> 通过将PC端获取得到的微信code去登录商城，本接口的Content-Type值为：application/x-www-form-urlencoded。因此请求的参数需要经过qs.stringify转换。baseURL备用地址(不在广州校区内网的，请统一使用这个地址)：

图 11-5　Content-Type 说明

需要在请求拦截器加上 **Content-Type** 请求头，代码如下：

```html
<!-- 第 11 章大型 PC 商城：单击"获取验证码"按钮的逻辑 -->
instance.interceptors.request.use(config=>{
    if (config.url === "/sendSMS" || config.url === "/wechatUsers/PCLogin") {
        config.headers["Content-Type"] = "application/x-www-form-urlencoded";
    }
    return config
},err=>{
    return Promise.reject(err)
})
```

下载 qs 模块，命令如下：

```
npm i qs
```

在 api.js 文件中导入 qs 模块，将参数转换为 queryString 格式，代码如下：

```js
import qs from "qs"
//发送短信验证码请求
export const SendSMSAPI = params =>
request.post("/sendSMS", qs.stringify(params));
```

再次到浏览器进行测试，看见响应为发送成功。

后端对每个手机号做了限制，如果看到如图 11-6 所示的提示，则说明已经达到最大可发送短信条数，请隔数小时后再尝试。

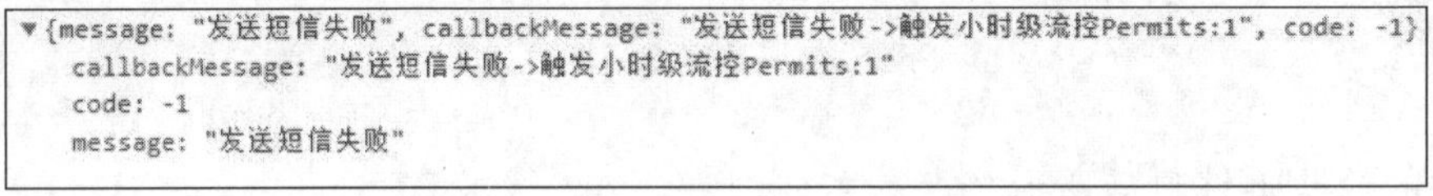

```
▼{message: "发送短信失败", callbackMessage: "发送短信失败->触发小时级流控Permits:1", code: -1}
  callbackMessage: "发送短信失败->触发小时级流控Permits:1"
  code: -1
  message: "发送短信失败"
```

图 11-6　手机号受小时限制响应图

### 11.7.7　请求成功回调函数的完善

请求成功后，显示请求成功提示，代码如下：

```html
<!-- 第 11 章大型 PC 商城：单击"获取验证码"按钮的逻辑 -->
//验证成功后，发起请求，如果获取验证码成功，则进行倒计时，并展示秒数
SendSMSAPI({
    phone:this.phoneNum.trim()
}).then(res=>{
    if(res.code===0){
        this.countdown();
```

```
            this.isShowCount=true;
            console.log(res);
        }else{
            //获取短信验证码失败
            alert(res.message)
        }
}).catch(err=>{
    //发送请求失败
    alert("请重新发送")
})
```

## 11.8 手机号码登录逻辑分析

手机号码登录逻辑分析：
（1）手机号码格式是否正确。
（2）拼图滑块验证通过。
（3）验证码是否为空（注意，这里前端并没有获得发送的手机验证码，所以只能判空）。
（4）发起登录请求。

### 11.8.1 抽取前两个验证的代码

前两个验证已经实现了，可以直接封装成函数。
在 Login.vue 文件中，封装验证方法，代码如下：

```
<!-- 第11章大型PC商城：手机号码登录逻辑分析 -->
toVerify(){
    //验证手机号是否正确
    if(!validateTelephoneNumber(this.phoneNum)){
        alert("请输入正确的手机号");
        this.$refs.phone.focus();
        return
    }

    //进行滑块验证
    if (this.msg == "再试一次" || this.msg == "向右滑动") {
        alert("先进行滑块验证");
        return
    }
},
getCode(){
    this.toVerify();
    //发起请求
    ...
}
```

如果只是按照上面的代码进行抽取，则会有验证不通过还发起请求的 Bug，所以将直接执行 this.toVerify()验证换成判断验证是否通过，如果不通过，则返回，如果验证成功，则发起请求，代码如下：

```
<!-- 第 11 章大型 PC 商城：手机号码登录逻辑分析 -->
toVerify(){
    //验证手机号是否正确
    if(!validateTelephoneNumber(this.phoneNum)){
        alert("请输入正确的手机号");
        this.$refs.phone.focus();
        return
    }

    //进行滑块验证
    if (this.msg == "再试一次" || this.msg == "向右滑动") {
        alert("先进行滑块验证");
        return
    }
    //返回验证通过
    return true
},
getCode(){
    if(!this.toVerify()){
        return
    };
    //验证成功后，发起请求
    ...
}
```

### 11.8.2  发起登录请求

在 api.js 文件中，添加手机登录请求方法，代码如下：

```
//手机号登录请求
export const PhoneLoginAPI = params =>
request.post("/phoneRegin",qs.stringify(params));
```

在 Login.vue 文件中，引入手机号登录请求方法，在登录按钮事件中，发起登录请求，代码如下：

```
<!-- 第 11 章大型 PC 商城：手机号码登录逻辑分析 -->
import { SendSMSAPI, PhoneLoginAPI } from "@/request/api";
...
        //单击"登录"按钮
        submitFn() {
            if(!this.toVerify()){
```

```
                return
            };

            //验证码是否为空
            if (this.code.trim() === "") {
                alert("请输入验证码再进行登录");
                return;
            }

            //发起登录请求
            PhoneLoginAPI({
                //先根据后端给的测试账号和密码进行登录
                //但在真正场景要传的是短信验证码和手机号
                phone:"13800138001",
                password:"qwerty567"
            }).then(res=>{
                //登录成功
                console.log(res);

            })
        },
```

登录成功，如图 11-7 所示。

```
▼ {code: 0, message: '登录成功!', x-auth-token: 'eyJ0eXAiOiJKV1QiLCJhbGciOiJI
    code: 0
    message: "登录成功!"
    x-auth-token: "eyJ0eXAiOiJKV1QiLCJhbGciOiJIUzI1NiJ9.eyJleHAiOjE2MzM0NTQwNz
  ▶ [[Prototype]]: Object
```

图 11-7　登录成功

## 11.8.3　登录成功后的逻辑

登录成功后，需要执行如下步骤。
（1）提示登录成功。
（2）将 token 值保存到 localStorage。
（3）隐藏登录模态窗口。
（4）登录状态的切换。
在回调函数中，提示登录成功、存储 token、隐藏登录模态窗，代码如下：

```
<!-- 第 11 章大型 PC 商城：手机号码登录逻辑分析 -->
            //发起登录请求
            PhoneLoginAPI({
                phone:"13800138001",
                password:"qwerty567"
            }).then(res=>{
                if(res.code===0){
```

```
            //提示登录成功
            alert("登录成功");

            //存储 token
            localStorage.setItem("x-auth-token", res["x-auth-token"]);

            //隐藏登录模态窗口
            this.chanIsShowLoginModal(false);

            //登录状态的切换
        }

    })
```

### 11.8.4 购物车按钮的布局

在 TopBar.vue 文件中,实现购物车按钮结构与布局,代码如下:

```
<!-- 第 11 章大型 PC 商城:手机号码登录逻辑分析 -->
<li class="cart_btn" v-if="isLogined">
    <img src="../assets/img/cart.png" alt="" width="20" />
    <span>购物车</span>
    <b>{{cartTotal}}</b>
</li>
<script>
    data () {
        return {
            //购物车总数
            cartTotal:0,
        }
    },
</script>

<style>
    .cart_btn {
        width: 124px;
        height: 40px;
        background: #0a328e;
        color: #fff;
        display: flex;
        justify-content: center;
        align-items: center;
        cursor: pointer;
        span {
            margin-left: 8px;
            margin-right: 6px;
        }
```

```
        b{
            width: 22px;
            height: 22px;
            line-height: 22px;
            border-radius: 50%;
            background-color: #f40;
            text-align: center;
        }
    }
</style>
```

### 11.8.5 购物车按钮展示（登录状态）分析

因为头部购物车按钮需要依靠用户是否登录这种状态的值进行展示，并且这个值会在 Login.vue 文件中进行修改，所以应把用户是否登录这种状态值放在 Vuex 中。

在 store 目录中新建文件夹 loginStatus，新建 index.js 文件，验证登录状态，代码如下：

```
<!-- 第 11 章大型 PC 商城：手机号码登录逻辑分析 -->
export default{
    namespaced:true,
    state: {
        isLogined:localStorage.getItem("x-auth-token")?true:false   //用来表
//示是否登录的登录状态值
    },
    mutations: {
        chanIsLogined(state,payload){
            console.log("执行了 chanIsLogined");
            state.isLogined = payload
        }
    }
}
```

在 store/index.js 文件中引入 loginStatus。

在 TopBar.vue 组件中，获取是否登录状态，代码如下：

```
<!-- 第 11 章大型 PC 商城：手机号码登录逻辑分析 -->
<div class="cart_btn" v-if="isLogined">
    <img src="../assets/img/cart.png" alt="" width="20" />
    <span>购物车</span>
    <b>{{cartTotal}}</b>
</div>
<li class="login-btn" v-else @click="chanIsShowLoginModal(true)">登录</li>

...
<script>
    import {mapMutations,mapState} from "vuex"
```

```
    ...
    computed:{
        ...mapState({
            isLogined:state=>state.loginStatus.isLogined
        })

    },
</script>
```

最后在登录的回调函数中修改这个值,即在 Login.vue 文件中将用户登录状态设置为 true,代码如下:

```
<!-- 第 11 章大型 PC 商城:手机号码登录逻辑分析 -->
...mapMutations({
        chanIsShowLoginModal:"showModal/chanIsShowLoginModal",
        chanIsLogined:"loginStatus/chanIsLogined"
    }),
...

    PhoneLoginAPI({
        ...
    }).then(res=>{
        if(res.code===0){
            //提示登录成功
            ...
            //存储 token
            ...
            //隐藏登录模态窗口
            ...
            //登录状态值的切换
            this.chanIsLogined(true)
        }
    })
```

## 11.9 提示组件的封装

### 11.9.1 icon 图标的使用

(1)在全局中引入 icon 图标,图标展示如图 11-8 所示。

图 11-8 中 icon 的链接为 https://at.alicdn.com/t/font_2730880_ylrio3ahhx.css。

图 11-8　icon 的展示

icon 图标及类名的展示如图 11-9 所示。

图 11-9　icon 图标及类名的展示

（2）具体图标名称如表 11-1 所示。

表 11-1　图标名称及类名

| 图 标 名 称 | 图 标 类 名 |
| --- | --- |
| YDUI 复选框（选中） | icon-yduifuxuankuangxuanzhong |
| YDUI 复选框 | icon-yduifuxuankuang |
| loading | icon-loading |
| toast 失败_画板 1 | icon-toast-shibai_huaban |
| toast 警告 | icon-toast-jinggao |
| toast_成功 | icon-toast_chenggong |

在组件中使用 iconfont，即给 HTML 元素添加 iconfont 相关类名，代码如下：

```
<i class="iconfont icon-loading"></i>
```

## 11.9.2　Toast 组件的初步封装与使用

图标样式网址为 https://at.alicdn.com/t/font_2730880_ylrio3ahhx.css。

将 iconfont 的样式链接内容粘贴到 src/assets/css/public.less 中，在 components 目录下新建 Toast.vue，实现 Toast 组件的基本结构与样式，代码如下：

```html
<!-- 第11章大型PC商城：提示组件的封装 -->
<template>
    <div class="toast">
        <i class="iconfont icon-toast-shibai_huaban"></i>
        <span>提示内容</span>
    </div>
</template>

<script>
export default {
   data () {
      return {


      }
   }
}
</script>

<style lang = "less" scoped>
.toast{
  position: fixed;
  padding: 10px 20px;
  display: flex;
  justify-content: center;
  align-items: center;
  background: #fff;
  left: 50%;
  top: 0;
  transform: translateX(-50%);
  border-radius: 10px;

  .iconfont{
    margin-right: 10px;
  }

  .icon-toast-shibai_huaban{
    color: red;
  }

  .icon-toast_chenggong{
    color: green;
  }

  .icon-toast-jinggao{
    color: orange;
```

```
    }
}
</style>
```

在 App.vue 文件中引入注册。

### 11.9.3　Toast 组件展示

Toast 组件展示与否最终可以在各个组件中调用，所以放在 Vuex 中，然后在 App 中，添加是否显示 Toast 组件，代码如下：

```
<!-- 第11章大型PC商城：提示组件的封装 -->
<Toast v-show="isShowToast"></Toast>
<script>
...
import {mapState} from "vuex"
export default {
    ...
    computed:{
        ...mapState({
            isShowToast:state=>state.showToast.isShowToast
        })
    }
}
</script>
```

接下来尝试通过事件验证是否能触发显示和隐藏切换。在 TopBar 组件中，先找头像尝试单击触发展示，代码如下：

```
<!-- 第11章大型PC商城：提示组件的封装 -->
<img @click="showToastFn" src="../assets/img/userImg.f8bbec5e.png" width="26" alt="">

<script>
    methods:{
        ...mapMutations({
            chanIsShowLoginModal:"showModal/chanIsShowLoginModal",
            chanIsShowToast:"showToast/chanIsShowToast"
        }),
        showToastFn(){
            this.chanIsShowToast(true)
        }
    }
</script>
```

### 11.9.4 Toast 组件的进场离场效果

Vue 提供了 transition 组件，配合 CSS3 可以用来实现进场离场效果，网址为 https://cn.vuejs.org/v2/guide/transitions.html。

在 App.vue 文件中添加进场与离场效果，代码如下：

```html
<!-- 第 11 章大型 PC 商城：提示组件的封装 -->
<template>
    <div id="app">
        <transition name="slide">
            <Toast v-show="isShowToast"></Toast>
        </transition>
        ...

    </div>
</template>
<style lang="less">
/* 入场的起始状态 = 离场的结束状态 */
.slide-enter, .slide-leave-to{
  opacity: 0;
}

.slide-enter-active, .slide-leave-active{
  transition: opacity .3s linear;
}

.slide-enter-to, .slide-leave{
  opacity: 1;
}
</style>
```

在 TopBar.vue 组件中的代码如下：

```js
methods:{
    ...mapMutations({
        chanIsShowLoginModal:"showModal/chanIsShowLoginModal",
        chanIsShowToast:"showToast/chanIsShowToast"
    }),
    showToastFn(){
        this.chanIsShowToast(true);
        setTimeout(()=>{
            this.chanIsShowToast(false);
        },1500)
    }
}
```

应把 Toast 组件中的 opacity 属性去掉。

## 11.9.5 封装 Toast 的属性

一个完整的 Toast 组件最好需要有展示、颜色、类型 3 种属性。在 Vuex 中补充相应属性，代码如下：

```
<!-- 第11章大型 PC 商城：提示组件的封装 -->
export default{
    namespaced:true,
    state: {
        //表示是否展示提示
        isShowToast:false,
        //toast 的内容
        toastMsg: "默认内容",
        //toast 的类型(success, danger, info)
        toastType: "success"
    },
    mutations: {
        chanIsShowToast(state,payload){
            console.log(payload);
            state.isShowToast = payload.isShow;
            if(payload.isShow){
                state.toastMsg = payload.msg;
                state.toastType = payload.type;
            }
        }
    },
    actions: {

    },
}
```

在 Toast.vue 组件中根据状态绑定相应样式，代码如下：

```
<!-- 第11章大型 PC 商城：提示组件的封装 -->
<template>
    <div class="toast">
        <!-- <i class="iconfont icon-toast-shibai_huaban"></i>
        <span>提示内容</span> -->
        <i
        :class="toastType=='success' ? 'iconfont icon-toast_chenggong' : (toastType=='danger' ? 'iconfont icon-toast-shibai_huaban' : 'iconfont icon-toast-jinggao')"
        ></i>
        <span>{{toastMsg}}</span>
    </div>
</template>
```

```
<script>
import {mapState} from "vuex"//引入 Vuex
export default {
  data () {
    return {


    }
  },
  computed:{
    ...mapState({
      toastMsg:state=>state.showToast.toastMsg,
      toastType:state=>state.showToast.toastType,
    })
  }
}
</script>
```

最后在 TopBar.vue 组件中调用时传入对象，代码如下：

```
<!-- 第 11 章大型 PC 商城：提示组件的封装 -->
showToastFn(){//显示 Toast
        this.chanIsShowToast({
            isShow:true,
            msg:"先登录",
            type:"danger"
        });
        setTimeout(()=>{
            this.chanIsShowToast({
              isShow:false,
            });
        },1500)
    }
```

### 11.9.6  Toast 组件自动关闭的处理

Toast 组件应该具备自动关闭功能，而不是每次调用都要写一段定时器代码来关闭，即在 TopBar.vue 组件中去掉 setTimeout 方法，代码如下：

```
<!-- 第 11 章大型 PC 商城：提示组件的封装 -->
 import {mapMutations,mapState,mapActions} from "vuex"

  methods:{
    ...mapMutations({
      chanIsShowLoginModal:"showModal/chanIsShowLoginModal"
    }),
    ...mapActions({
```

```
        asyncIsShowToast:"showToast/asyncIsShowToast"
    }),
    showToastFn(){
        this.asyncIsShowToast({
            isShow:true,
            msg:"先登录",
            type:"danger"
        });
        /* setTimeout(()=>{
            this.chanIsShowToast({
                isShow:false,
            });
        },1500) */
    }
}
```

在 Vuex 中书写 actions,代码如下:

```
<!-- 第 11 章大型 PC 商城:提示组件的封装 -->
actions: {
    asyncIsShowToast(context,payload){//异步显示 Toast 组件
        context.commit("chanIsShowToast",payload)
        setTimeout(()=>{
            context.commit("chanIsShowToast",{
                isShow:false
            })
        },1000)
    }
}
```

## 11.9.7　总结:提示框组件的使用

以后在任意组件中都可使用提示框,代码如下:

```
<!-- 第 11 章大型 PC 商城:提示组件的封装 -->
import {mapActions} from "vuex"

    methods:{
        ...mapActions({
            asyncChanToastState:"showToast/asyncChanToastState"
        }),
        showToastFn(){
            this.asyncChanToastState({
                msg:"先登录",
                type:"danger"
            });
        }
    }
}
```

## 11.10 微信扫码登录——微信登录二维码的获取与展示

### 11.10.1 获取微信二维码

在 public/index.html 文件的 head 标签中引入，代码如下：

```
<script src="https://res.wx.qq.com/connect/zh_CN/htmledition/js/wxLogin.js"></script>
```

把 Login.vue 文件中展示二维码图片的盒子上添加 id="weixin"，获取的微信二维码会以 iframe 的方式嵌入这个盒子中，代码如下：

```
<div id="weixin" class="qrcode" v-show="!isShowForm">
    二维码
</div>
```

单击切换到微信登录的函数中，代码如下：

```
<!-- 第11章大型PC商城：微信扫码登录——微信登录二维码的获取与展示 -->
weixinClick(){//单击切换微信扫码登录这一项，并向微信扫码登录
        this.isShowForm=false;

        //微信登录第1步：申请微信登录二维码
        let _this = this;
        new WxLogin({
            id: "weixin",
            appid: "wx67cfaf9e3ad31a0d",
            scope: "snsapi_login",
            //扫码成功后重定向的接口
            redirect_uri: "https://sc.wolfcode.cn/cms/wechatUsers/shop/PC",
            //state 填写编码后的URL
            state:encodeURIComponent(window.btoa("http://127.0.0.1:8080"+_this.$route.path)),
            //调用样式文件
            href: "",
        });
    },
```

### 11.10.2 微信二维码样式调整

把 wxLoginStyle 文件夹放到 utils 文件夹中，然后在这个 wxLoginStyle 目录下用 node 执行 JS 文件，命令如下：

```
node data-url.js
```

得到的信息如下：

```
data:text/css;base64,Lyogd3hsb2dpbi5jc3MgKi8NCi5pbXBvd2VyQm94IC50aXRsZSSwg
LmltcG93ZXJCb3ggLmluZm97DQogIGRpc3BsYXk6IG5vbmU7DQp9DQoNCi5pbXBvd2VyQm94IC5xc
mNvZGV7DQogIG1hcmdpbi10b3A6IDIwcHg7DQp9
```

把它填到上面的 href 属性中，隐藏头部尾部。

最后调整页面 iframe 外层盒子的样式，使二维码居中，代码如下：

```
<!-- 第 11 章大型 PC 商城：微信扫码登录——微信登录二维码的获取与展示 -->
#weixin{
    /* background-color: #fcf; */
    display: flex;
    justify-content: center;
    margin-top: -20px;
}
```

至此已经可以进行扫码登录了。

如果扫码跳转时被 Chrome 浏览器拦截，则需要修改 Chrome 浏览器设置。打开 Chrome 浏览器设置，搜索"弹出式窗口和重定向"，如图 11-10 所示。

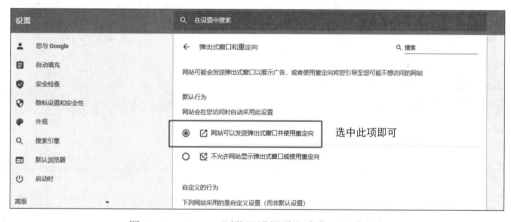

图 11-10　Chrome 浏览器设置弹出式窗口和重定向

## 11.11　微信扫码登录——用临时票据 code 换取 token

需要在顶部组件 TopBar.vue 加载时用临时票据 code 换取 token，代码如下：

```
<!-- 第 11 章大型 PC 商城：微信扫码登录——用临时票据 code 换取 token -->
import {WeixinLoginApi} from "@/request/api"

...
created(){
    //微信登录第 2 步：用临时票据 code 换取 token
    let mycode = this.$route.query.code;
```

```
      if (mycode) {
        //有 code 才去换取 token
        WeixinLoginApi({
          code: mycode,
        }).then((res) => {
          console.log(res);
          if (res.code === 0) {
            //登录成功
            //提示用户登录成功
            this.asyncChanToastState({
              msg:"登录成功",
              type:"success"
            });
            //保存 token 值
            localStorage.setItem("x-auth-token", res["x-auth-token"]);
            //改变登录状态
            this.chanIsLogined(true);
          }
        })
      }
    }
```

登录成功后，在成功的回调函数中清除网址栏上显示的 code，代码如下：

```
<!-- 第 11 章大型 PC 商城：微信扫码登录——用临时票据 code 换取 token -->
//清除浏览器网址栏上的 code
this.$router.push(this.$route.path);
/*
setTimeout(()=>{
   this.$router.go(0)
},2000)
*/
```

当返回的 res.code 不为 0 时，表示登录失败，代码如下：

```
<!-- 第 11 章大型 PC 商城：微信扫码登录——用临时票据 code 换取 token -->
if (res.code === 0) {
   ...
}else{
   //当登录失败时给出提示信息
   this.asyncChanToastState({
      msg:res.message,
      type:"danger"
   });
}
```

当无法获取 code 时，也更新一下用户登录状态，代码如下：

```
<!-- 第 11 章大型 PC 商城：微信扫码登录——用临时票据 code 换取 token -->
```

```
if(mycode){
    ....
}
else{
    //当没有mycode时也设置用户状态

    //如果没有code，则说明用户没扫码
    //说明用户已经登录的方式不用扫码，或者用户没有登录
    //判断用户能否获得token来更新登录状态
    let mytoken = localStorage.getItem("x-auth-token")
    this.chanIsLogined(Boolean(mytoken));
}
```

## 11.12 手机验证码登录

在 Login.vue 文件中，修改两个参数，并且补上 res.code==0，但是每天一个手机号有限定发送短信验证码的次数，代码如下：

```
<!-- 第11章大型PC商城：手机验证码登录 -->
//去做登录
    LoginAPI({
      phone: this.phoneNumber,
      password: this.SMScode,
    }).then((res) => {
      console.log(res);
      if(res.code==0){

        alert("登录成功");
        localStorage.setItem("x-auth-token", res["x-auth-token"]);
        this.chanIsShowLoginMadal(false);
        this.chanIsLogined(true);

      }else{
        alert(res.message);
      }
    });
```

## 11.13 路由监听及其应用

在登录状态下，如果手动删除 token，则在切换路由时购物车依然存在。这是因为登录状态依旧没有改变。

### 1. 路由监听的格式

在 TopBar.vue 文件中书写 watch 监听属性，监听 $route.path 的变化，可以检测是否有路由跳转，代码如下：

```
<!-- 第11章大型PC商城：路由监听及其应用 -->
watch: {
    //监听路由的变化
    "$route.path": {
      handler(newVal, oldVal){
          //什么时候执行这里的代码？当路由发生变化时执行
          console.log(newVal, oldVal);
      }
    }
},
```

### 2. 路由监听的应用

可以把 TopBar.vue 文件中 created 的代码封装起来，看作设置用户登录状态的函数，封装在 methods 中，代码如下：

```
<!-- 第11章大型PC商城：路由监听及其应用 -->
created(){
   this.setUserLoginstatus()
},
watch: {
    //监听路由的变化
    "$route.path": {
        handler(newVal, oldVal){
            //console.log(newVal, oldVal)
            if(newVal !== oldVal){
              this.setUserLoginstatus()
            }
        }
    }
},
methods:{
   setUserLoginstatus(){
       //微信登录第2步：用临时票据code换取token
       let mycode = this.$route.query.code;
        if (mycode) {
           ...
        }else{
           ...
        }
    },
}
```

## 11.14 组件重载

之前已经实现了当完成切换路由时就更新用户登录状态值功能。这个功能，也可以使用组件重载的方式来完成，即每次切换路由都让 TopBar.vue 重新加载一次。

首先，在 TopBar 的 created()函数中补充更新用户状态值的代码，删除 watch 中对路由的监听，代码如下：

```
<!-- 第 11 章大型 PC 商城：组件重载 -->
    if(mycode){
      ...
    }else{
      //如果不是通过微信登录的，就会执行这里的代码
      let mytoken = localStorage.getItem("x-auth-token");
      this.chanIsLogined(Boolean(mytoken));
    }
```

其次，利用 key 属性实现组件重载。来到 App.vue 组件，给调用 TopBar.vue 的地方添加 key 属性，并监听 key 属性值的变化，实现组件重载，代码如下：

```
<!-- 第 11 章大型 PC 商城：组件重载 -->
<template>
  ...
  <!-- 顶部 -->
  <TopBar :key="topBarKeyValue"></TopBar>
  ...
</template>
<script>
  data() {
    return {
      topBarKeyValue:1
    };
  },
  ...
  watch: {
    //监听路由的变化
    "$route.path": {
      handler(newVal, oldVal){
        //console.log(newVal, oldVal)
        if(newVal !== oldVal){
          //key 属性的值一变化，就会执行组件重载操作，从而执行 created()函数
          console.log("组件重载！");
          this.topBarKeyValue++;
        }
      }
    }
  }
```

```
    },
</script>
```

## 11.15 获取登录用户信息

**1. 请求头携带 token**

在登录功能中，目前已经完成了将 token 保存到本地存储中。

在真正项目中，只要本地存储中有 token，在请求时都会在每个请求头中带上这个 token 值，不管这个请求需不需要 token 都会带上。接下来，就需要在请求拦截器中判断 token，以及携带 token。

在 src/request/request.js 文件中获取 token，如果 token 存在，则在请求头中添加 x-auth-token 属性，代码如下：

```
<!-- 第11章大型PC商城：获取登录用户信息，请求头携带 token -->
instance.interceptors.request.use(config => {
    const token = localStorage.getItem("x-auth-token");
    if (token) {
      //判断是否存在token，如果存在，则每个请求的请求头上都加上token
      config.headers["x-auth-token"] = token;
    }
    return config
}, err => {
    return Promise.reject(err)
})
```

**2. 获取用户登录信息**

接口文档网址为 http://www.docway.net/project/1h9xcTeAZzV/1hG2hFlipBQ?st=1iUU09vKhMm&sid=1iUU09vvKhM。

在 api.js 文件中定义获取用户信息方法，代码如下：

```
//获取登录用户信息
export const UserProfilesAPI = () => request.get("/shop/userProfiles");
```

在 TopBar 组件中，引入获取用户信息方法，并在 setUserLoginstatus 方法实现用户信息的获取，代码如下：

```
<!-- 第11章大型PC商城：获取登录用户信息，请求头携带 token -->
import {UserProfilesAPI} from "@/request/api"
setUserLoginstatus(){
     //微信登录第2步：用临时票据code换取token
     let mycode = this.$route.query.code;
     if (mycode) {
        ...
     }else{
```

```js
            //如果没有code，则说明没有扫码，或者用户已经登录了，但不用扫码
            //判断有没有token,设置登录状态（因为TopBar上面的信息需要靠有没有登录去展示的）
            let mytoken = localStorage.getItem("x-auth-token");
            this.chanIsLogined(Boolean(mytoken));
            if(mytoken){
                //请求并渲染用户信息
                UserProfilesAPI().then(res=>{
                    //打印用户信息
                    console.log(res);
                });
            }else{
                //设置回默认的用户信息
            }
        }
    },
```

## 11.16　用户信息渲染

为了在项目中实现任意组件都可以获取用户信息，例如 TopBar.vue、Login.vue 等，可以将用户信息放在 Vuex 中。在 store 中新建 userInfo 文件夹，新建 index.js 文件，添加用户信息数据与获取用户信息方法，代码如下：

```js
<!-- 第11章大型PC商城：用户信息渲染 -->
import {UserProfilesAPI} from "@/request/api"

export default {
    namespaced: true,
    state: {
        //购物车数量
        cartTotal: 0,
        //用户信息
        userInfo:{
            headImg:require("../../assets/img/userImg.f8bbec5e.png"),
            nickName:"--",
            coin:"--"
        }
    },
    mutations: {
        updateUserInfo(state,payload){//更新用户信息
            console.log("payload 为",payload);
            state.cartTotal = payload.cartTotal;
            state.userInfo = payload.userInfo;
        }
    },
    actions: {
```

```
        asyncUpdateUserInfo(context){//异步更新用户信息
            UserProfilesAPI().then(res=>{
                //打印用户信息
                context.commit("updateUserInfo",res.data)
            });
        }
    },
}
```

在 **TopBar.vue** 组件中实现用户信息的获取，代码如下：

```
<!-- 第 11 章大型 PC 商城：用户信息渲染 -->
<template>
  ...
  <li>
      <img
          @click="clickAvatar"
          class="avatar"
          width="26"
          :src="userInfo.headImg"
          alt=""
          />用户名：{{userInfo.nickName}}
  </li>
  <li>我的鸡腿：{{userInfo.coin}}</li>
  <li>获取鸡腿</li>
  <li>叩丁狼官网</li>
  <li class="cart-btn btn" v-show="isLogined">
      <img src="../assets/img/cart.png" alt="" />
      <span>购物车</span>
      <b>{{ cartTotal }}</b>
  </li>
  ...
</template>
<script>
    setUserLoginstatus(){
        //微信登录第 2 步：用临时票据 code 换取 token
        let mycode = this.$route.query.code;
        if (mycode) {
            //有 code 才去换取 token
            WeixinLoginApi({
                code: mycode,
            }).then((res) => {
                console.log(res);
                if (res.code === 0) {
                    //登录成功后：提示用户登录成功；保存 token 值；改变登录状态；清除浏览器网
//址栏上的 code；获取登录用户信息
                    this.asyncUpdateUserInfo();
```

```
            }else{
              this.asyncChanToastState({
                msg:res.message,
                type:"danger"
              });
            }
          })
        }else{
          //如果没有code，则可通过能否获得存储的token来更新登录状态
          let mytoken = localStorage.getItem("x-auth-token");
          this.chanIsLogined(Boolean(mytoken));
          //如果mytoken存在，则获取登录用户信息
          if(mytoken){
              this.asyncUpdateUserInfo();
          }
        }
      },
    }
</script>
```

在 Login.vue 组件中如果登录成功，则异步更新用户信息，代码如下：

```
<!-- 第 11 章大型 PC 商城：用户信息渲染 -->
import { mapMutations,mapActions } from "vuex";
...
  ...mapActions({
    asyncUpdateUserInfo:"userInfo/asyncUpdateUserInfo"
  }),
  toLogin() {
    //前两个验证已经完成
    //判断验证码是否为空
    //去登录
    LoginAPI({
      ...
    }).then((res) => {
      console.log(res);
      if(res.code==0){
          //提示登录成功；保存token值，本地存储；隐藏登录框；登录状态的切换；更新用
//户信息
          this.asyncUpdateUserInfo();
      }
    });
  },
```

如果此时用户头像没有出来，则在 public 文件夹中的 index.html 文件的 head 标签里添加这个标签即可，代码如下：

```
<!-- 防止服务器检查防盗用链接 -->
<meta name="referrer" content="no-referrer" />
```

## 11.17　删除 token 后的用户信息初始化

在删除 token 后，切换路由时，用户信息需要还原为默认值。
在 TopBar.vue 文件中，执行初始化用户数据方法，代码如下：

```
<!-- 第 11 章大型 PC 商城：删除 token 后的用户信息初始化 -->
setUserLoginState() {
   if(mycode){
      ...
   }else {
      ...
      if (mytoken) {
         ...
      } else {
         //初始化用户数据
         this.initUserInfo();
      }
   }
}
```

## 11.18　首页布局的套用

在实际工作项目中，有可能项目部分代码并不是由你完成的，而是由你的同事完成的。甚至有可能在你进入公司之后，才接手同事的项目，这时就要求你看得懂别人写的代码。

接下来套用首页的结构和样式，把以下提供的首页文件内容替换到项目对应的文件（src/views/Home.vue）中，然后在首页文件的基础上进行数据渲染。

src/views/Home.vue 文件中的代码如下：

```
<!-- 第 11 章大型 PC 商城：首页布局的套用 -->
<template>
  <div class="home">
    <div class="banner wrap">
     <img src="../assets/img/banner.f559b49d.png" alt="" />
    </div>
    <div class="content">
     <div class="wrap">
        <JfTitle title1="精品推荐" :imgSrc="titImg1"></JfTitle>
        <List :arr="jingpinArr"></List>
        <JfTitle title1="热门兑换" :imgSrc="titImg2"></JfTitle>
        <img
          style="margin: 10px 0 30px 0"
          src="../assets/img/ad.4c6b6225.png"
          alt=""
        />
```

```vue
      <List :arr="remenArr"></List>
    </div>
  </div>
  <div class="wrap">
    <JfTitle title1="积分攻略" :imgSrc="titImg3"></JfTitle>
    <ul class="jifen">
      <li :style="{ backgroundImage: `url(${jifenImg1})` }">
        <h3>签到得鸡腿</h3>
        <div>去签到</div>
      </li>
      <li :style="{ backgroundImage: `url(${jifenImg2})` }">
        <h3>购课得鸡腿</h3>
        <div>去购课</div>
      </li>
      <li :style="{ backgroundImage: `url(${jifenImg3})` }">
        <h3>推荐得鸡腿</h3>
        <div>去推荐</div>
      </li>
      <li :style="{ backgroundImage: `url(${jifenImg4})` }">
        <h3>做任务得鸡腿</h3>
        <div>做任务</div>
      </li>
    </ul>
  </div>
</div>
</template>

<script>
//@ is an alias to /src
import JfTitle from "@/components/home/JfTitle";
import List from "@/components/home/List";
export default {
  name: "Home",
  components: { JfTitle, List },
  data() {
    return {
      titImg1: require("../assets/img/jingpin.png"),
      titImg2: require("../assets/img/hot.png"),
      titImg3: require("../assets/img/score.png"),
      jifenImg1: require("../assets/img/integral-01.9386d4bf.png"),
      jifenImg2: require("../assets/img/integral-02.150d92a1.png"),
      jifenImg3: require("../assets/img/integral-03.9870f3f1.png"),
      jifenImg4: require("../assets/img/integral-04.afadcbdf.png"),
      jingpinArr: [],
      remenArr: [],
    };
  },
```

```
    created() {

    },
};
</script>

<style lang="less" scoped>
.banner {
  padding-bottom: 30px;
}
.content {
  background-color: #f5f5f5;
  padding-bottom: 30px;
}
.jifen {
  display: flex;
  justify-content: space-between;
  padding-bottom: 30px;
  li {
    width: 285px;
    height: 168px;
    color: #fff;
    padding: 20px 10px;
    box-sizing: border-box;
    cursor: pointer;
    transition: background-size 0.4s linear;
    background-size: 100% 100%;
    &:hover {
      background-size: 105% 105%;
    }
    h3 {
      font-size: 24px;
      margin-bottom: 19px;
    }
    div {
      width: 96px;
      height: 27px;
      border: 1px solid #ffffff;
      text-align: center;
      line-height: 27px;
    }
  }
}
</style>
```

src/components/Footer.vue 文件中的代码如下：

```html
<!-- 第 11 章大型 PC 商城：首页布局的套用 -->
<template>
  <div class="footer-b">
    <div class="heart wrap">
      <div class="footer-i">
        <h2>
          <a href=""><img src="../assets/img/slogan.7730f7f2.png" alt="" />
          </a>
        </h2>
        <p style="line-height: 1.6em">
          叩丁狼是一家专注于培养高级 IT 技术人才，为学员提供定制化 IT 职业规划方案及意见咨询服务的教育科技公司，为学员提供海量优质课程，以及创新的线上线下学习体验，帮助学员获得全新的个人发展和能力提升。
        </p>
      </div>
      <ul>
        <li>
          <img src="../assets/img/wx.5584e874.png" alt="" />
          <a href="javascript:;">微信公众号</a>
        </li>
        <li><a href="javascript:;">新浪微博</a></li>
        <li>
          <img src="../assets/img/service.848ec511.png" alt="" />
          <a href="javascript:;">在线客服</a>
        </li>
      </ul>
      <div class="footer-r">
        <p>全国免费咨询热线</p>
        <h3>020-85628002</h3>
      </div>
    </div>
  </div>
</template>

<script>
export default {
  data() {
    return {};
  },
};
</script>

<style lang = "less" scoped>
.footer-b {
  width: 100%;
  /* margin-top: 20px; */
  height: 240px;
```

```
        background: #242b39;
        .heart {
          display: flex;
          justify-content: space-between;
          .footer-i {
            padding-top: 32px;
            p {
              padding-top: 27px;
              width: 426px;
              height: 55px;
              font-size: 12px;
              font-family: Microsoft YaHei;
              font-weight: 400;
              color: #7d879a;
            }
          }
          ul {
            margin-top: 123px;
            width: 293px;
            display: flex;
            justify-content: space-between;
            li {
              border-right: 1px solid #9e9e9e;
              height: 20px;
              position: relative;
              text-align: center;
              img {
                position: absolute;
                left: -20%;
                top: -110px;
                display: none;
              }
              &:hover {
                img {
                  display: block;
                }
              }
              a {
                height: 19px;
                font-size: 16px;
                font-family: Microsoft YaHei;
                font-weight: 400;
                color: #7d879a;
                padding-right: 16px;
                text-decoration: none;
              }
              &:last-child {
```

```
          border-right: 0;
        }
      }
    }
    .footer-r {
      padding-top: 90px;
      p {
        width: 117px;
        height: 14px;
        font-size: 14px;
        font-family: Microsoft YaHei;
        font-weight: 400;
        color: #7d879a;
      }
      h3 {
        margin-top: 10px;
        width: 220px;
        height: 22px;
        font-size: 28px;
        font-family: SourceHanSansSC;
        font-weight: bold;
        color: #ffffff;
      }
    }
  }
}
</style>
```

src/components/home/JfTitle.vue 文件中的代码如下：

```
<!-- 第 11 章大型 PC 商城：首页布局的套用 -->
<template>
  <div class="title">
    <div class="l">
      <img :src="imgSrc" alt="" />
      <h2>{{ title1 }}</h2>
    </div>
    <div class="r" v-show="title1 != '积分攻略'">
      更多
      <span><img src="../../assets/img/arrow.png" alt="" /></span>
    </div>
  </div>
</template>

<script>
export default {
  props: ["title1", "imgSrc"],
```

```
  data() {
    return {};
  },
};
</script>

<style lang = "less" scoped>
.title {
  display: flex;
  justify-content: space-between;
  padding-top: 50px;
  padding-bottom: 20px;
  .l {
    display: flex;
    align-items: center;
    font-size: 30px;
    font-family: SourceHanSansSC;
    font-weight: bold;
    color: #242b39;
    h2 {
      margin-left: 10px;
    }
  }
  .r {
    display: flex;
    align-items: center;
    cursor: pointer;
    span {
      margin-left: 6px;
    }
  }
}
</style>
```

src/components/home/List.vue 文件中的代码如下：

```
<!-- 第 11 章大型 PC 商城：首页布局的套用 -->
<template>
  <ul class="list">
    <li v-for="item in [10,20,30,40]" :key="item">
      <section>
        <img src="../../assets/img/listimg.jpg" alt="" />
        <div class="bottom-box">
          <h3>商品标题</h3>
          <p>200 积分</p>
          <div class="btn">立即兑换</div>
        </div>
```

```html
        <img
          class="flag"
          src="../../assets/img/section_new.png"
          alt=""
        />
        <img
          class="flag"
          src="../../assets/img/section_hot.png"
          alt=""
        />
      </section>
    </li>
  </ul>
</template>

<script>
export default {
  data() {
    return {};
  },
  methods: {

  },
};
</script>

<style lang = "less" scoped>
.list {
  /* width: 1220px; */
  /* overflow: hidden; */
  display: flex;
  /* 自动换行 */
  flex-wrap: wrap;

  li {
    &:nth-of-type(4n) {
      margin-right: 0px;
    }
    width: 285px;
    float: left;
    margin-bottom: 20px;
    margin-right: 20px;
    position: relative;
    section {
      width: 285px;
      height: 492px;
      position: relative;
```

```css
      transition: all 0.5s linear;
    }
    .bottom-box {
      width: 285px;
      height: 162px;
      background-color: #fff;
      text-align: center;
      h3 {
        font-size: 18px;
        font-family: SourceHanSansSC;
        font-weight: 300;
        color: #333333;
        height: 56px;
        line-height: 56px;
        /* background-color: #ffc; */
      }
      p {
        font-size: 18px;
        font-family: SourceHanSansSC;
        font-weight: bold;
        color: #fd604d;
      }
      .btn {
        width: 100px;
        height: 36px;
        line-height: 36px;
        border: 1px solid #0a328e;
        margin: 20px auto 0;
        font-size: 18px;
        font-family: SourceHanSansSC;
        font-weight: 300;
        color: #0a328e;
      }
    }
    .flag {
      position: absolute;
      left: 0;
      top: 0;
    }
    &:hover {
      cursor: pointer;
      section {
        margin-top: -15px;
        box-shadow: 0 0 10px #ccc;
```

```css
      .btn {
        background-color: #0a328e;
        color: #fff;
      }
    }
  }
}
</style>
```

在 api.js 文件中，添加精品推荐模块与热门兑换模块数据请求方法，代码如下：

```js
//首页精品推荐请求
export const JingpinAPI = () => request.get("/products/recommend");
//请求热门兑换的数据
export const RemenAPI = () => request.get("/products/hot");
```

在 src/views/Home.vue 文件中，执行精品推荐模块与热门兑换模块数据请求，代码如下：

```js
<!-- 第 11 章大型 PC 商城：首页布局的套用 -->
async created() {
    //精品请求
    let res1 = await JingpinAPI();
    //过滤出前 4 项
    if (res1.code == 0) {
        this.jingpinArr = res1.data.data.records.filter(
            (item, index) => index < 4
        );
    }

    //热门请求
    let res2 = await RemenAPI();
    if (res2.code == 0) {
        this.remenArr = res2.data.data.records.filter((item, index) => index < 6);
    };
    console.log(this.jingpinArr,this.remenArr);
},
```

在 src/components/home/List.vue 文件中，实现数据渲染，代码如下：

```html
<li v-for="item in arr" :key="item.id">
    <section>
        <img :src="imgBaseUrl + item.coverImg" alt="" />
        <div class="bottom-box">
            <h3>{{ item.name }}</h3>
            <p>{{ item.coin }}积分</p>
            <div class="btn">立即兑换</div>
        </div>
    </section>
```

```html
            <img
                class="flag"
                v-show="item.isLatest == 1"
                src="../../assets/img/section_new.png"
                alt=""
            />
            <img
                class="flag"
                v-show="item.isHotSale == 1"
                src="../../assets/img/section_hot.png"
                alt=""
            />
    </section>
</li>
```

在 main.js 文件中保存这个基本图片的 URL，代码如下：

```js
//在组件内部获取这个变量
//像一个"全局变量"，在任意的组件中都可以通过 this.变量名或者{{变量名}} 获取
//但并不像 Vuex 那样共享数据，而仅仅是组件内部的一个变量
Vue.prototype.imgBaseUrl = "http://sc.wolfcode.cn";
```

## 11.19　详情页的处理

### 1. 单击商品跳转到详情页

在 views 下新建详情页组件 Details.vue，并在路由文件中添加路由配置，代码如下：

```js
<!-- 第 11 章大型 PC 商城：详情页的处理 -->
{
    path: '/details',
    name: 'Details',
    component: () => import(/* webpackChunkName: "details" */ '../views/Details.vue')
}
```

在 List 组件的 li 上添加单击跳转事件，并携带参数，代码如下：

```html
<!-- 第 11 章大型 PC 商城：详情页的处理 -->
<li v-for="item in arr" :key="item.id" @click="goToDetails(item.id)">
    ...
</li>
<script>
export default {
    ...
    methods:{
        goToDetails(id){
            this.$router.push(`/details?id=${id}`);
```

        }
      }
    };
</script>
```

在 Details.vue 组件中可以获取这个 id 值,代码如下:

```
<!-- 第11章大型PC商城:详情页的处理 -->
<template>

        这里是详情页{{$route.query.id}}

</template>

<script>
export default {
    data () {
        return {

        }
    },
    created(){
        console.log(this.$route.query.id);
    }
}
</script>
```

### 2. 单击商品跳转到详情页

在 api.js 文件中添加请求详情页数据的方法,代码如下:

```
//请求详情页的数据
export const GoodDetailsAPI = (id) => request.get(`/products/${id}`);
```

在 Details.vue 组件中,发起请求获取数据,代码如下:

```
<!-- 第11章大型PC商城:详情页的处理 -->
<template>

        这里是详情页{{$route.query.id}}

</template>

<script>
import {GoodDetailsAPI} from "@/request/api"
export default {
    created(){
        let goodId = this.$route.query.id;
        //获取id后,发起请求
```
```

```
          GoodDetailsAPI(goodId).then(res=>{
              console.log(res);
          })
      }
}
</script>
```

### 3. 详情页结构样式

先在 components 文件夹中新建面包屑组件 Crumb.vue，将提供的其他文件夹中的结构样式填入 Details.vue 文件中，图片也放到对应的文件夹中。

### 4. 解构请求到的数据，渲染右侧模块

在 Details.vue 文件中解构请求到的数据，代码如下：

```
<!-- 第11章大型PC商城：详情页的处理 -->
<script>
data () {
    return {
        ...
        //面包屑数据
        nav:[],
        //产品数据
        productInfo:{},
        //还可以兑换
        themYouCanBuy:[]
    }
},
created(){
    let goodId = this.$route.query.id;
    GoodDetailsAPI(goodId).then(res=>{
        if(res.code==0){
            let {nav,productInfo,themYouCanBuy } = res.data
            this.nav = nav;
            this.productInfo = productInfo;
            this.themYouCanBuy = themYouCanBuy;
        }
    })
}
</script>
```

对右侧"还可以兑换"模块进行渲染，代码如下：

```
<!-- 第11章大型PC商城：详情页的处理 -->
<h3>还可以兑换</h3>
<ul>
    <li v-for="item in themYouCanBuy" :key="item.id">
        <div class="l">
            <img :src="`https://sc.wolfcode.cn`+item.img" alt="">
```

```
            </div>
            <div class="r">
                <div class="title">{{item.name}}</div>
                <div class="score">
                    <span>{{item.coin}}</span>
                    积分
                </div>
            </div>
        </li>
</ul>
```

#### 5. 抽取图片基本路径

通过 Vue.prototype.变量名=值定义的变量在组件内部可以直接用 this.变量名获取。有人称它为全局变量，但它不是真正意义上的全局变量。跟 Vuex 也不一样。它相当于在每个组件中都定义了一个组件内部变量。

在 main.js 文件中添加图片服务器地址，代码如下：

```
//定义"全局变量"，相当于在每个组件中都定义了一个组件内部变量 imgBaseUrl
Vue.prototype.imgBaseUrl = "https://sc.wolfcode.cn";
```

在 Details.vue 和 List.vue 组件中，图片基本路径的处理均调整为图片服务器地址+具体的图片地址，代码如下：

```
<img :src="imgBaseUrl+item.img" alt="">
```

#### 6. 单击跳转对应的详情页

单击"还可以兑换"后跳转到对应的详情页，代码如下：

```
<!-- 第 11 章大型 PC 商城：详情页的处理 -->
<li v-for="item in themYouCanBuy" :key="item.id" @click="$router.push(`/details?id=${item.id}`)">

    ...
<script>
  watch: {
    /* "$route": {
      handler(newVal, oldVal){
        console.log(newVal, oldVal)
      },
      deep: true
    }, */
    "$route.query.id": {
      handler(newVal, oldVal) {
        if (newVal !== oldVal) {
          //刷新当前页
          this.$router.go(0);
        }
```

## 7. 左侧大图和小图的切换渲染

具体的代码如下：

```
<!-- 第 11 章大型 PC 商城：详情页的处理 -->
<div class="bigImg">
        <img
            :src="imgBaseUrl + productInfo.coverImg"
            width="100%"
        />
    </div>
    <ul>
        <li v-for="(item, index) in productInfo.imgAltas"
          :key="index"
          @mouseenter="imgTab(item.src,index)">
            <img
                :src="imgBaseUrl + item.src"
                width="100"
                :style="{opacity:(imgTabIndex==index)?1:0.5}"
            />
        </li>
    </ul>
<script>
    data(){
        ...
        //图片切换栏的当前 index
        imgTabIndex:0
    },
    methods:{
        imgTab(src,i){
            this.productInfo.coverImg=src;
            this.imgTabIndex=i;
        }
    },
</script>
```

## 8. 选择颜色的渲染

对 Sku 模块中的"选择颜色"进行渲染，代码如下：

```
<!-- 第 11 章大型 PC 商城：详情页的处理 -->
<section v-for="(item, index) in productInfo.parameterJson"
        :key="index">
    <strong>选择{{item.title}}</strong>
    <ul>
        <li
```

```
            :class="val.currentActivate ? 'active' : ''"
            v-for="val in item.parameters"
            :key="val.id"
            @click="$router.push(`/details?id=${val.id}`)"
          >
            {{ val.title }}
          </li>
      </ul>
</section>
<script>
    ...
    data(){
        //产品数据
        productInfo: {
            parameterJson: [
                {
                    parameters: [],
                },
            ],
        },
    }
</script>
```

9. 步进器

步进器的代码如下：

```
<!-- 第11章大型PC商城：详情页的处理 -->
<div class="step">
        <div class="reduce" @click="stepFn(-1)">-</div>
        <input type="text" disabled v-model="stepNum" />
        <div class="add"  @click="stepFn(1)">+</div>
      </div>
<script>
    ...
    stepFn(val){
        if(this.stepNum-1==0 &&val==-1){
            return
        }
        this.stepNum+=val
    }
</script>
```

10. 面包屑渲染

在 Crumb.vue 组件中添加具名插槽，代码如下：

```
<slot name="slot1"></slot> /
<slot name="slot1"></slot>
```

在 Details.vue 组件中渲染面包屑组件，代码如下：

```
<Crumb>
<span slot="slot1">{{nav[0].name}}</span>
<span slot="slot1">{{nav[1].name}}</span>
</Crumb>
```

### 11. 面包屑优化

上面这种方式将插槽和组件写死，万一 nav 是多层级的，则需要优化。

在 Details.vue 组件中，优化面包屑的渲染，代码如下：

```
<!-- 第11章大型PC商城：详情页的处理 -->
<span
    v-for="(item, index) in nav"
    :key="index"
    v-text="index < nav.length - 1 ? item.name + ' / ' : item.name"
    >
    <!-- {{index==nav.length-1 ? item.name : item.name + '/' }} -->
</span>
```

在 Crumb.vue 组件中，修改插槽，代码如下：

```
<slot></slot>
```

### 12. 礼品详情及切换栏

礼品详情及切换栏部分的代码如下：

```
<!-- 第11章大型PC商城：详情页的处理 -->
<ul class="tabs">
        <li @click="flag = true" :class="flag === true ? 'active' : ''">
            礼品详情
        </li>
        <li @click="flag = false" :class="flag === false ? 'active' : ''">
            常见问题
        </li>
    </ul>
<div v-show="flag" v-html="productInfo.description"></div>
<div v-show="!flag" class="issue">...</div>
<script>
    ...
    data(){
        //详情和问题的切换
        flag:true
    },
    created(){
      ...
        GoodDetailsAPI(goodId).then(res=>{
            if(res.code==0){
```

```
            ...
            this.productInfo = productInfo;
            //解决图片无法显示的问题
            this.productInfo.description =
this.productInfo.description.replace(
                /upload/g,
                this.imgBaseUrl+"/upload"
            );
            ...
        }
    })
}
</script>
```

## 11.20 单击加入购物车

接口网址为 http://112.124.4.201/project/1h9xcTeAZzV/1hJt0rsYfsO?st=1iUU09vKhMm&sid=1iUU09vKhMv。

在 api.js 文件中，添加加入购物车功能的方法，代码如下：

```
//加入购物车
export const AddToCartApi = (params) => request.post(`/shop/carts/add`,
qs.stringify(params));
```

在 Details.vue 文件中发起加入购物车功能请求，代码如下：

```
<!-- 第11章大型PC商城：单击加入购物车 -->
<div class="addToCart" @click="addToCart">加入购物车</div>
<script>
    import {AddToCartApi} from "@/request/api"
    ...
        addToCart(){
            AddToCartApi({
                productId: this.$route.query.id,
                total: this.stepNum,
                modified: 1,
            }).then((res) => {
                if (res.code === 0) {
                    //提示请求成功
                    this.asyncChanIsShowToast({
                        msg: res.message,
                        type: "success",
                    });
                    //更新购物车数字，子传父，通知父级更新Header组件
                    this.$emit("fn");
                    /* 或者直接刷新页面  this.$router.go(0) */
```

```
                    }
                });
            }
</script>
```

App.vue 文件中的代码如下:

```
<!-- 第 11 章大型 PC 商城:单击加入购物车 -->
<router-view @fn="reloadTopBar"/>

<script>
    methods:{
        reloadTopBar(){
            this.num++
        }
    },
</script>
```

**注意** 购物车商品的总数,本项目中后端计算的逻辑是商品种类,而不是总件数。另外,加入购物车,也只是替换件数,而不是增加件数。

## 11.21 全部商品页面

### 11.21.1 结构样式套用

在 views 目录中新建 Goods.vue 组件,代码如下:

```
<!-- 第 11 章大型 PC 商城:全部商品页面 -->
<template>
<div class="goods">
  <div class="wrap">
    <Crumb></Crumb>
    <img src="../assets/img/banner.4c6b6225.png" width="100%" alt="" />
    <ul class="options">
      <li>
        <strong>排序:</strong>
        <span class="active">全部</span>
        <span>我还可以兑换</span>
        <span>0-500 个</span>
        <span>500-1000 个</span>
        <span>1000-1500 个</span>
        <span>1500-2500 个</span>
      </li>
      <li>
        <strong>分类:</strong>
```

```html
          <span  class="active">全部</span>
          <span>实物礼品</span>
          <span>虚拟礼品</span>
        </li>
      </ul>
      <List :arr="[10,20,30]" />
    </div>
  </div>
</template>
<script>

import Crumb from "../components/Crumb.vue";
import List from "../components/home/List.vue";
export default {
  data() {
    return {
    };
  },
  components: {
    Crumb,
    List,
  },

};
</script>

<style lang = "less" scoped>
@import "../assets/css/public.less";

.goods {
  padding-bottom: 50px;
  background: #efefef;
  padding-top: 20px;
  .options {
    padding-top: 20px;
    padding-bottom: 40px;
    li {
      margin-top: 20px;
      strong {
        color: #000;
        font-weight: bold;
      }
      span {
        margin-right:20px;
        margin-left: 10px;
        cursor: pointer;
        color: #999;
```

```
        &.active {
          color: @base-color;
          font-weight: bold;
        }
      }
    }
  }
}
</style>
```

### 11.21.2 商品列表渲染

接口地址为 http://112.124.4.201/project/1h9xcTeAZzV/1hGAyV68R9c?st=1iUU09vKhMm&sid=1iUU09vKhMm。

在 API 中添加商品搜索方法，代码如下：

```
//商品搜索（含首页的"更多"）
export const GoodsSearchApi = (params) => request.get(`/products`, {params});
```

在 Goods.vue 文件中定义与商品搜索数据相关的参数，实现获取商品搜索数据功能，代码如下：

```
<!-- 第 11 章大型 PC 商城：全部商品页面 -->
<script>
    import {GoodsSearchApi} from "@/request/api"
    ....
  data() {
      return {

          //栏目目前可提供的参数是 1->精品推荐；2->热门兑换；0->全部
          did: 0,
          //( 1->实体商品；2->虚拟商品；0->全部 )
          type: 0,
          //大于多少积分，最少是 0
          min: 0,
          //少于多少积分，最多是 10000，当传入 0 时，会直接返回所有商品，无视后台逻辑
          max: 0,
          //商品关键词
          keyword: "",
          //商品列表数组
          goodsList: [],
      };
    },
    created(){
        GoodsSearchApi({
            did: this.did,
```

```
              type: this.type,
              min: this.min,
              max: this.max,
              keyword: this.keyword,
          }).then((res) => {
              if (res.code === 0) {
                this.goodsList = res.data;
              }
          });
      },
</script>
```

### 11.21.3 选项数据的分析和渲染

选项部分可以看作下拉菜单的选择，所以选用 options 数组。
在 data 中补充数据，代码如下：

```
<!-- 第 11 章大型 PC 商城：全部商品页面 -->
//排序的数组项
    options1: [
      { min: 0, max: 10000, txt: "全部" },
      //"我还可以兑换"的 max 值是 Header 组件的 coin 值
      { min: 0, max: 0, txt: "我还可以兑换" },
      { min: 0, max: 500, txt: "0~500 分" },
      { min: 500, max: 1000, txt: "500~1000 分" },
      { min: 1000, max: 1500, txt: "1000~1500 分" },
      { min: 1500, max: 2500, txt: "1500~2500 分" },
      { min: 2500, max: 6500, txt: "2500~6500 分" },
      { min: 6500, max: 10000, txt: "6500~10000 分" },
    ],
    //分类数组项
    options2: [
      { type: 0, txt: "全部" },
      { type: 1, txt: "实体商品" },
      { type: 2, txt: "虚拟商品" },
    ],
    //排序当前项
    num1:0,
    //分类当前项
    num2:0,
```

因此修改结构，代码如下：

```
<!-- 第 11 章大型 PC 商城：全部商品页面 -->
<li>
        <strong>排序：</strong>
        <span
```

```
            :class="num1==index?'active':''"
            v-for="(item, index) in options1"
            :key="index"
            @click="num1=index"
        >{{ item.txt }}</span>
    >
  </li>
  <li>
      <strong>分类:</strong>
      <span
         :class="num2==index?'active':''"
         v-for="(item, index) in options2"
         :key="index"
         @click="num2=index"
      >{{ item.txt }}</span>
    >
  </li>
```

## 11.21.4 单击选项，切换商品列表

点选每项，处理当前样式都应该发起刚才的 GoodsSearchApi 请求，所以把它封装到 methods 中，代码如下：

```
<!-- 第 11 章大型 PC 商城：全部商品页面 -->
@click="options1Click(index,item.min,item.max)"
...
@click="options2Click(index,item.type)"
...
<script>
created(){
    this.goodsSearch();
},
methods:{
    options1Click(index,min,max){
        //当前样式
        this.num1=index;
        this.min=min
        this.max=max
        this.goodsSearch()
    },
    options2Click(index,type){
        //当前样式
        this.num2=index;
        this.type=type;
        this.goodsSearch()
    },
```

```
    goodsSearch(){
        GoodsSearchApi({
            did: this.did,
            type: this.type,
            min: this.min,
            max: this.max,
            keyword: this.keyword,
        }).then((res) => {
            if (res.code === 0) {
                this.goodsList = res.data;
            }
        });
    }
  },
</script>
```

### 11.21.5 搜索框事件

在 Header.vue 组件中，添加搜索框事件，代码如下：

```
<!-- 第11章大型PC商城：全部商品页面 -->
<input @keyup.13="toSearch" type="text" placeholder="请输入要搜索的商品" v-model="userInputSearch"/>
<span @click="toSearch"><img src="../assets/img/search.png" alt="" /></span>
<script>
export default {
  data() {
    return {
      //用户输入的搜索内容
      userInputSearch:""
    };
  },
  methods:{
    //执行搜索
    toSearch(){
      //跳转到goods页面并携带参数
      this.$router.push(`/goods?keyword=${this.userInputSearch}`);
      //清空内容
      this.userInputSearch = "";
    }
  }
};
</script>
```

此时，搜索虽然能够跳转，但是对应的商品列表请求还缺少关键字。

在 Goods.vue 组件中，获取搜索框传入的搜索关键字，代码如下：

```
created(){
    this.keyword = this.$route.query.keyword||""
    this.goodsSearch()
},
```

但是,目前只能在首页搜索跳转 goods 页面,goods 页面搜索不会跳转,所以还需要添加路由监听功能,以便刷新当前页,代码如下:

```
<!-- 第 11 章大型 PC 商城:全部商品页面 -->
watch: {
    "$route.query.keyword": {
        handler(newVal, oldVal) {
            this.keyword = newVal
            this.goodsSearch();
        },
    }
},
```

## 11.22 导航守卫

前置导航守卫有以下 3 个参数。
(1) to:表示即将进入的路由。
(2) from:表示即将离开的路由。
(3) next():表示执行进入这个路由。

### 11.22.1 全局导航守卫

在 router/index.js 文件中,添加路由的全局导航守卫,代码如下:

```
<!-- 第 11 章大型 PC 商城:导航守卫(导航拦截、路由拦截) -->
//全局导航守卫
router.beforeEach((to, from, next)=>{
    //如果有 token 就表示已经登录
    //如果想要进入个人中心页面,则必须有登录标识 token

    //console.log('to:', to)
    //console.log('from:', from)
    if(to.path=='/user'){
        let token = localStorage.getItem('x-auth-token')
        //此时必须有 token
        if(token){
            next(); //next()到 to 所对应的路由界面
        }else{
            //提示没有登录
            store.dispatch("showToast/asyncChanIsShowToast",{
```

```
                msg: "你还没有登录!",
                type: "danger",
            })
        }
        return; //需要return, 防止执行完上面的next()后还继续执行下面的next()
    }
    //如果不是去往个人中心的路由, 则直接通过守卫转到to所对应的路由界面
    next()
})

export default router
```

### 11.22.2 组件内部导航守卫

全局导航守卫在每次改变路由时都会触发, 然而, 目前只有到个人中心页面才触发, 所以需要在组件内部书写导航守卫。在User.vue组件中, 实现组件内部导航守卫, 代码如下:

```
<!-- 第11章大型PC商城: 导航守卫(导航拦截、路由拦截) -->
import store from "@/store"

...
beforeRouteEnter(to, from, next) {
    //在渲染该组件的对应路由被 confirm 前调用
    //不能获取组件实例 this
    //因为当守卫执行前, 组件实例还没被创建
    let token = localStorage.getItem("x-auth-token");
    if(token){
      next()
    }else{
      //提示没有登录
      store.dispatch("showToast/asyncChanIsShowToast",{
        msg: "你还没有登录!",
        type: "danger",
      });
    
    }
},
```

## 11.23 个人中心——购物车页面

**1. 个人中心左侧结构样式套用**

把以下的User.vue组件代码套用到项目views目录中:

```
<!-- 第11章大型PC商城: 个人中心——购物车页面 -->
```

```html
<template>
  <div class="person_page">
    <div class="person_wrap">
      <Crumb :nav="[{name:'首页'},{name:'个人中心'}]"></Crumb>
      <main>
        <aside>
          <!-- <div
            class="avatar"
            :style="{ backgroundImage: `url(${userInfo.headImg})` }"
          ></div>
          <div class="name">{{ userInfo.nickName }} <span @click="loginOutFn" style="cursor:pointer;">[退出]</span></div> -->
          <div class="title">
            <img
              src="../assets/img/transaction.png"
              width="20"
              alt="交易管理"
            />
            交易管理
          </div>
          <ul class="list">
            <li :class="/\/person1/g.test($route.path) ? 'active' : ''">
              个人中心
            </li>
            <li :class="/\/person1/g.test($route.path) ? 'active' : ''">
              我的订单
            </li>
            <li :class="/\/cart/g.test($route.path) ? 'active' : ''">购物车</li>
            <li :class="/\/person1/g.test($route.path) ? 'active' : ''">
              消息通知
            </li>
            <li :class="/\/person1/g.test($route.path) ? 'active' : ''">
              积分明细
            </li>
            <li :class="/\/person1/g.test($route.path) ? 'active' : ''">
              积分攻略
            </li>
          </ul>
          <div class="title">
            <img
              src="../assets/img/transaction.png"
              width="20"
              alt="交易管理"
            />
            个人信息管理
          </div>
```

```html
        <ul class="list">
          <li>地址管理</li>
          <li>账号安全</li>
        </ul>
      </aside>
      <article><router-view></router-view></article>
    </main>
  </div>
</div>
</template>
<script>
import Crumb from "@/components/Crumb.vue";

import store from "@/store"
export default {
  data() {
    return {

    };
  },
  components: {
    Crumb,
  },
  methods: {
    loginOutFn(){
      localStorage.removeItem("x-auth-token");
      setTimeout(()=>{
        this.$router.push('/home');
      }, 2000)
    }
  },
  beforeRouteEnter(to, from, next) {
    //在渲染该组件的对应路由被 confirm 前调用
    //不能获取组件实例 this
    //因为当守卫执行前，组件实例还没被创建
    let token = localStorage.getItem("x-auth-token");
    if(token){
      next()
    }else{
      //提示没有登录
      store._actions["toastStatus/asyncChanIsShowToast"][0]({
        msg: "你还没有登录！2222",
        type: "danger",
      });


    }
  },
```

```less
};
</script>
<style lang="less" scoped>
@import "../assets/css/public.less";
.person_page {
  background: #fff;
  main {
    border-top: 1px solid #e1e1e1;
    padding: 28px 0 48px;
    display: flex;
    justify-content: space-between;
    background: #fff;
    aside {
      width: 200px;
      height: 740px;
      background: #e7e7e7;
      margin-right: 60px;
      box-sizing: border-box;
      padding: 30px 18px 0;
      .avatar {
        width: 100px;
        height: 100px;
        margin: auto;
        background-size: 100% 100%;
        background-repeat: no-repeat;
      }
      .name {
        text-align: center;
        margin-top: 19px;
        margin-bottom: 43px;
        span {
          text-decoration: underline;
          color: #2a5df1;
        }
      }
      .title {
        font-size: 16px;
        color: #333333;
        display: flex;
        align-items: center;
        margin-bottom: 14px;
        img {
          margin-right: 6px;
        }
      }
      .list {
        li {
```

```
            margin-bottom: 17px;
            font-weight: 300;
            color: #666666;
            cursor: pointer;
            &.active {
              color: @base-color;
              font-weight: bold;
              &::before {
                width: 2px;
                height: 14px;
                background:@base-color;
                display: inline-block;
                content: "";
                margin-right: 10px;
              }
            }
          }
        }
      }
      article {
        flex: 1;
        padding: 20px 0 0 0px;
        box-sizing: border-box;
        background: #fff;
      }
    }
  }
</style>
```

处理头像和用户名的展示，代码如下：

```
<!-- 第11章大型 PC 商城：个人中心——购物车页面 -->
import {mapState} from "vuex";
...
computed:{
   ...mapState({
      userInfo:state=>state.userInfo.userInfo
   })
},
```

### 2. 重定向到购物车及其结构样式套用

需求：一进入个人中心页面，就跳转到购物车界面。

分析：购物车是个人中心众多页面下的一个，会把它做成子路由。

先准备购物车组件：在 components 下新建 user 文件夹，新建 cart.vue 组件。

在 router/index.js 下配置重定向到 /user 的子路由，代码如下：

```
<!-- 第11章大型 PC 商城：个人中心——购物车页面 -->
```

```
{
   path: '/user',
   name: 'User',
   component: () => import(/* webpackChunkName: "user" */ '../views/User.vue'),
   redirect:"/user/cart",
   children:[
     {
       path: 'cart',
       component: () => import(/* webpackChunkName: "cart" */ '../components/user/Cart.vue'),
     }
   ]
},
```

直接把以下 Cart.vue 套用到项目 components 目录中，代码如下：

```
<!-- 第 11 章大型 PC 商城：个人中心——购物车页面 -->
<template>
  <div class="cart_page">
    <table>
      <thead>
        <tr>
          <th style="width: 8%">
            <i
              :class="
                checkAll
                  ? 'iconfont icon-yduifuxuankuangxuanzhong'
                  : 'iconfont icon-yduifuxuankuang'
              "
            ></i>
          </th>
          <th style="width: 30%">礼品信息</th>
          <th>兑换分数</th>
          <th>数量</th>
          <th>小计（鸡腿）</th>
          <th>操作</th>
        </tr>
      </thead>
      <tbody>
        <tr>
          <td>
            <i
              :class="
                //checkList[index]
                  ? 'iconfont icon-yduifuxuankuangxuanzhong'
                  : 'iconfont icon-yduifuxuankuang'
```

```
              "
            ></i>
          </td>
          <td>
            <section>
              <img
                width="84"
                src="http://sc.wolfcode.cn/upload/images/product_images/20200615/41ddc8c8-bd4b-4f5c-ae68-474f9ed18eb7.png"
                alt="列表图片"
              />
              <div class="info">
                <h5>叮丁狼定制T恤</h5>
                <p>颜色、版本:XL</p>
              </div>
            </section>
          </td>
          <td>5000 鸡腿</td>
          <td>
            <div class="step">
              <span>-</span>
              <input type="text" disabled v-model="stepNum" />
              <span>+</span>
            </div>
          </td>
          <td>5000 鸡腿</td>
          <td>
            <span class="del">删除</span>
          </td>
        </tr>
      </tbody>
    </table>
    <div class="total">总计：<span>0 鸡腿</span></div>
    <div class="submit">提交</div>
  </div>
</template>
<script>
export default {
  data() {
    return {
      stepNum: 1,
      //全选
      checkAll: false,
      //单选按钮数组
      checkList: [],
    };
  },
```

```less
};
</script>
<style lang="less" scoped>
.cart_page {
  background: #fff;
  table {
    width: 100%;
    border: 1px solid #e6e6e6;
    box-sizing: border-box;
    color: #666;
    border-collapse: collapse;
    font-size: 14px;
    thead {
      background: #f2f2f2;
      th {
        padding: 19px 0;
        .iconfont {
          cursor: pointer;
        }
        .icon-yduifuxuankuangxuanzhong {
          color: #0a328e;
        }
      }
    }
    tbody {
      tr {
        td {
          vertical-align: middle;
          text-align: center;
          padding: 19px 0;
          table-layout: fixed; //td的宽度固定，不随内容变化
          .iconfont {
            cursor: pointer;
          }
          .icon-yduifuxuankuangxuanzhong {
            color: #0a328e;
          }
          section {
            padding-left: 20px;
            display: flex;
            box-sizing: border-box;
            img {
              margin-right: 12px;
            }
            .info {
              padding-top: 20px;
              flex: 1;
```

```css
      overflow: hidden;
      box-sizing: border-box;
      text-align: left;
      h5 {
        overflow: hidden;
        color: #333;
        font-size: 18px;
        white-space: nowrap;
        text-overflow: ellipsis;
        margin-bottom: 20px;
      }
      p {
        color: #666;
        overflow: hidden;
        white-space: nowrap;
        text-overflow: ellipsis;
      }
    }
  }
  .step {
    width: 106px;
    height: 32px;
    margin: auto;
    span {
      float: left;
      width: 30px;
      height: 32px;
      display: block;
      border: solid 1px #d1d1d1;
      font-size: 20px;
      box-sizing: border-box;
      font-weight: normal;
      font-stretch: normal;
      line-height: 30px;
      letter-spacing: 0px;
      color: #999999;
      text-align: center;
      cursor: pointer;
      background: #fff;
    }
    input {
      box-sizing: border-box;
      width: 46px;
      height: 32px;
      float: left;
      text-align: center;
      font-size: 14px;
```

```
            line-height: 23px;
            letter-spacing: 0px;
            color: #666666;
            border: 0;
            border-top: 1px solid #d1d1d1;
            border-bottom: 1px solid #d1d1d1;
            background: #fff;
          }
        }
        .del {
          border: 1px solid #ececec;
          border-radius: 4px;
          padding: 5px 10px;
          cursor: pointer;
          &:hover {
            color: #fff;
            background: #0a328e;
          }
        }
      }
    }
  }
}
.total {
  padding: 30px 0;
  text-align: right;
  font-size: 22px;
  span {
    font-weight: bold;
    color: #fd604d;
  }
}
.submit {
  width: 175px;
  height: 40px;
  text-align: center;
  line-height: 40px;
  font-family: SourceHanSansSC-Light;
  font-size: 18px;
  font-weight: normal;
  font-stretch: normal;
  letter-spacing: 0px;
  color: #ffffff;
  cursor: pointer;
  background-color: #0a328e;
  float: right;
}
```

```
}
</style>
```

### 3. 购物车数据请求及渲染

在 api.js 文件中添加获取购物车数据方法,代码如下:

```
//请求购物车数据
export const CartDataApi = () => request.get(`/shop/carts`);
```

在 cart.vue 组件中获取购物车数据,代码如下:

```
<!-- 第11章大型PC商城：个人中心——购物车页面 -->
<td>
        <section>
          <img
            width="84"
            :src="imgBaseUrl+item.coverImg"
            alt="列表图片"
          />
          <div class="info">
            <h5>{{item.title}}</h5>
            <p>{{item.versionDescription}}</p>
          </div>
        </section>
      </td>
      <td>{{item.coin}}鸡腿</td>
      <td>
      <div class="step">
        <span>-</span>
        <input type="text" disabled v-model="item.total" />
        <span>+</span>
      </div>
      </td>
      <td>{{item.coin * item.total}}鸡腿</td>
      <td>
        <span class="del">删除</span>
      </td>
<script>
import {CartDataApi} from "@/request/api";
export default {
  data() {
    return {
      ...
      //购物车数组
      cartArr:[]
    };
  },
  created(){
```

```
        CartDataApi().then(res=>{
            if(res.code===0){
                this.cartArr = res.data
            }
        })
    }
};
</script>
```

## 11.24  404 处理

新建 Error.vue 组件，实现 404 页面结构，代码如下：

```
<!-- 第 11 章大型 PC 商城：404 处理 -->
<template>
  <div class="wrap"><img width="100%" src="../assets/img/404.94e7c552.jpg" alt=""></div>
</template>
```

将 Error.vue 添加到路由列表中，代码如下：

```
<!-- 第 11 章大型 PC 商城：404 处理 -->
  {
    path: '*',
    name: 'Error',
    component: () => import(/* webpackChunkName: "error" */ '../views/Error.vue')
  }
```

## 11.25  滚动到底部加载更多

在 Goods.vue 文件中，添加加载模块，代码如下：

```
<p style="text-align: center; margin-top: 20px">
    正在加载... ...
</p>
```

三个函数（带兼容性的写法）的代码如下：

```
<!-- 第 11 章大型 PC 商城：滚动到底部加载更多 -->
//获取滚动条当前的位置
function getScrollTop() {
    var scrollTop = 0;
    if(document.documentElement && document.documentElement.scrollTop) {
        scrollTop = document.documentElement.scrollTop;
    } else if(document.body) {
```

```
        scrollTop = document.body.scrollTop;
    }
    return scrollTop;
}

//获取当前可视范围的高度
function getClientHeight() {
    var clientHeight = 0;
    if(document.body.clientHeight && document.documentElement.clientHeight) {
        clientHeight = Math.min(document.body.clientHeight,
document.documentElement.clientHeight);
    } else {
        clientHeight = Math.max(document.body.clientHeight,
document.documentElement.clientHeight);
    }
    return clientHeight;
}

//获取文档完整的高度
function getScrollHeight() {
    return Math.max(document.body.scrollHeight,
document.documentElement.scrollHeight);
}
```

在 Goods.vue 文件中实现数据加载，代码如下：

```
<!-- 第 11 章大型 PC 商城：滚动到底部加载更多 -->
<script>
import {getScrollTop,getClientHeight,getScrollHeight} from "@/utils"
...
data(){
    return{
        //产品列表
        goodsList: [],
        //用来展示的产品列表
        goodsListShow: [],
        //默认展示第 1 页
        page:1,
        //每页 8 条
        size:8,
        //false 表示没有正在加载
        isLoading: false,
        //是否已经到底
        isReachBottom:false,

    }
},
```

```js
...
getGoodList() {
  GoodListAPI({
    ...
  }).then((res) => {
    ...
    //在请求到数据之后,先展示前 8 条
    this.goodsListShow=this.goodsList.filter((item,index)=>index<8)
  });
},
scrollFn() {
  //如果滚动就执行这里的代码,频繁地触发事件
  console.log(getClientHeight() + getScrollTop() == getScrollHeight() + 1);
  //console.log("页面正在滚动");
  //如果滚动到底部
  //if (到底部了) {
  //if (窗口高度+scrollTop>=页面文档高度-20) {
  if (getClientHeight() + getScrollTop() >= getScrollHeight() - 20) {

    //需要 this.isLoading 为 false 时才能进行加载
    if (!this.isLoading) {
      //this.isLoading 避免了重复触发这个到底了加载数据事件
      this.page++;
      this.isLoading = true;
      setTimeout(() => {
        //往 goodsListShow 数组 push 下一页的数据
        //从 goodsList 数组中将 this.page 页的数据 push 到 goodsListShow
        for (var i = this.size * (this.page - 1); i < this.size * this.page; i++) {
          //this.goodsList[i]必须有这个值,才能 push 到展示的数组里
          this.goodsList[i]
            ? this.goodsListShow.push(this.goodsList[i])
            : "";
        }
        this.isLoading = false;
      }, 500);
    }
  }
},
},
mounted() {
  //监听滚动
  window.addEventListener("scroll", this.scrollFn);
},
beforeDestroy() {
  //取消监听
  window.removeEventListener("scroll", this.scrollFn);
},
```

处理是否已经到底部 isReachBottom，结构完善后的代码如下：

```html
<!-- 第 11 章大型 PC 商城：滚动到底部加载更多 -->
<p style="text-align: center; margin-top: 20px">
    {{ isReachBottom ? "已经没有数据了" : "正在加载……" }}
</p>
<script>
...
//定义是不是已经没有数据了
isReachBottom: false,
  ...
    if (getClientHeight() + getScrollTop() >= getScrollHeight() - 20) {
        if (this.goodsListShow.length >= this.goodsList.length) {
            //没有数据了
            this.isReachBottom = true;
            return;
        }
</script>
```

最后解决选项切换时的 Bug，代码如下：

```
async goodsSearch(){
    ...
    //初始化数据
    this.isReachBottom = false;
    this.page = 1;

    //判断是不是已经没有数据了
    if (this.goodsListShow.length >= this.goodsList.length) {
      //每次请求到数据，把页数和是否到底部初始化一下
      this.isReachBottom = true;
    }
}
```

## 11.26 跨域配置

对 vue.config.js 进行跨域配置，代码如下：

```
<!-- 第 11 章大型 PC 商城：跨域配置 -->
module.exports = {
   devServer: {
      proxy: {
         '/api': {
            target: "http://kumanxuan1.f3322.net:8881/cms",
            pathRewrite: {
```

```
                    '^/api': ''
                }
            }
        }
    }
}
```

在 request.js 文件中,配置请求前缀,代码如下:

```
const instance = axios.create({
    baseURL: "/api",
    timeout: 5000
})
```

配置完后需要重启服务器。

## 11.27 项目环境变量配置

项目目录下新建两个文件,分别用于存放开发环境和生产环境下的不同配置。
在 .env.dev 中,配置开发环境,代码如下:

```
NODE_ENV=development
VUE_APP_BASE_URL=http://192.168.113.249:8081/cms
VUE_APP_STATE_URL=http://127.0.0.1:8080
```

在 .env.prod 中,配置生产环境,代码如下:

```
NODE_ENV = production
VUE_APP_BASE_URL = http://kumanxuan1.f3322.net:8881/cms
VUE_APP_STATE_URL="后端给的地址"
```

在 package.json 文件中修改启动命令,修改后的命令如下:

```
"serve": "Vue-CLI-service serve --open --mode dev",
"servepro": "Vue-CLI-service serve --open --mode prod",
```

在 vue.config.js 文件中将服务器地址替换为配置文件中的地址,代码如下:

```
'/api': {
    target: process.env.VUE_APP_BASE_URL,
    pathRewrite: {
        '^/api': ''
    }
}
```

# 图 书 推 荐

| 书 名 | 作 者 |
|---|---|
| 深度探索 Vue.js——原理剖析与实战应用 | 张云鹏 |
| 剑指大前端全栈工程师 | 贾志杰、史广、赵东彦 |
| Flink 原理深入与编程实战——Scala+Java（微课视频版） | 辛立伟 |
| Spark 原理深入与编程实战（微课视频版） | 辛立伟、张帆、张会娟 |
| PySpark 原理深入与编程实战（微课视频版） | 辛立伟、辛雨桐 |
| HarmonyOS 移动应用开发（ArkTS 版） | 刘安战、余雨萍、陈争艳 等 |
| HarmonyOS 应用开发实战（JavaScript 版） | 徐礼文 |
| HarmonyOS 原子化服务卡片原理与实战 | 李洋 |
| 鸿蒙操作系统开发入门经典 | 徐礼文 |
| 鸿蒙应用程序开发 | 董昱 |
| 鸿蒙操作系统应用开发实践 | 陈美汝、郑森文、武延军、吴敬征 |
| HarmonyOS 移动应用开发 | 刘安战、余雨萍、李勇军 等 |
| HarmonyOS App 开发从 0 到 1 | 张诏添、李凯杰 |
| HarmonyOS 从入门到精通 40 例 | 戈帅 |
| JavaScript 基础语法详解 | 张旭乾 |
| 华为方舟编译器之美——基于开源代码的架构分析与实现 | 史宁宁 |
| Android Runtime 源码解析 | 史宁宁 |
| 鲲鹏架构入门与实战 | 张磊 |
| 鲲鹏开发套件应用快速入门 | 张磊 |
| 华为 HCIA 路由与交换技术实战 | 江礼教 |
| 华为 HCIP 路由与交换技术实战 | 江礼教 |
| openEuler 操作系统管理入门 | 陈争艳、刘安战、贾玉祥 等 |
| 恶意代码逆向分析基础详解 | 刘晓阳 |
| 深度探索 Go 语言——对象模型与 runtime 的原理、特性及应用 | 封幼林 |
| 深入理解 Go 语言 | 刘丹冰 |
| Spring Boot 3.0 开发实战 | 李西明、陈立为 |
| 深度探索 Flutter——企业应用开发实战 | 赵龙 |
| Flutter 组件精讲与实战 | 赵龙 |
| Flutter 组件详解与实战 | [加]王浩然（Bradley Wang） |
| Flutter 跨平台移动开发实战 | 董运成 |
| Dart 语言实战——基于 Flutter 框架的程序开发（第 2 版） | 亢少军 |
| Dart 语言实战——基于 Angular 框架的 Web 开发 | 刘仕文 |
| IntelliJ IDEA 软件开发与应用 | 乔国辉 |
| Vue+Spring Boot 前后端分离开发实战 | 贾志杰 |
| Vue.js 快速入门与深入实战 | 杨世文 |
| Vue.js 企业开发实战 | 千锋教育高教产品研发部 |
| Python 从入门到全栈开发 | 钱超 |
| Python 全栈开发——基础入门 | 夏正东 |
| Python 全栈开发——高阶编程 | 夏正东 |
| Python 全栈开发——数据分析 | 夏正东 |
| Python 编程与科学计算（微课视频版） | 李志远、黄化人、姚明菊 等 |
| Python 游戏编程项目开发实战 | 李志远 |
| 量子人工智能 | 金贤敏、胡俊杰 |
| Python 人工智能——原理、实践及应用 | 杨博雄 主编，于营、肖衡、潘玉霞、高华玲、梁志勇 副主编 |
| Python 预测分析与机器学习 | 王沁晨 |

续表

| 书 名 | 作 者 |
|---|---|
| Python 数据分析实战——从 Excel 轻松入门 Pandas | 曾贤志 |
| Python 概率统计 | 李爽 |
| Python 数据分析从 0 到 1 | 邓立文、俞心宇、牛瑶 |
| FFmpeg 入门详解——音视频原理及应用 | 梅会东 |
| FFmpeg 入门详解——SDK 二次开发与直播美颜原理及应用 | 梅会东 |
| FFmpeg 入门详解——流媒体直播原理及应用 | 梅会东 |
| FFmpeg 入门详解——命令行与音视频特效原理及应用 | 梅会东 |
| Python Web 数据分析可视化——基于 Django 框架的开发实战 | 韩伟、赵盼 |
| Python 玩转数学问题——轻松学习 NumPy、SciPy 和 Matplotlib | 张骞 |
| Pandas 通关实战 | 黄福星 |
| 深入浅出 Power Query M 语言 | 黄福星 |
| 深入浅出 DAX——Excel Power Pivot 和 Power BI 高效数据分析 | 黄福星 |
| 云原生开发实践 | 高尚衡 |
| 云计算管理配置与实战 | 杨昌家 |
| 虚拟化 KVM 极速入门 | 陈涛 |
| 虚拟化 KVM 进阶实践 | 陈涛 |
| 边缘计算 | 方娟、陆帅冰 |
| 物联网——嵌入式开发实战 | 连志安 |
| 动手学推荐系统——基于 PyTorch 的算法实现（微课视频版） | 於方仁 |
| 人工智能算法——原理、技巧及应用 | 韩龙、张娜、汝洪芳 |
| 跟我一起学机器学习 | 王成、黄晓辉 |
| 深度强化学习理论与实践 | 龙强、章胜 |
| 自然语言处理——原理、方法与应用 | 王志立、雷鹏斌、吴宇凡 |
| TensorFlow 计算机视觉原理与实战 | 欧阳鹏程、任浩然 |
| 计算机视觉——基于 OpenCV 与 TensorFlow 的深度学习方法 | 余海林、翟中华 |
| 深度学习——理论、方法与 PyTorch 实践 | 翟中华、孟翔宇 |
| HuggingFace 自然语言处理详解——基于 BERT 中文模型的任务实战 | 李福林 |
| Java+OpenCV 高效入门 | 姚利民 |
| AR Foundation 增强现实开发实战（ARKit 版） | 汪祥春 |
| AR Foundation 增强现实开发实战（ARCore 版） | 汪祥春 |
| ARKit 原生开发入门精粹——RealityKit + Swift + SwiftUI | 汪祥春 |
| HoloLens 2 开发入门精要——基于 Unity 和 MRTK | 汪祥春 |
| 巧学易用单片机——从零基础入门到项目实战 | 王良升 |
| Altium Designer 20 PCB 设计实战（视频微课版） | 白军杰 |
| Cadence 高速 PCB 设计——基于手机高阶板的案例分析与实现 | 李卫国、张彬、林超文 |
| Octave 程序设计 | 于红博 |
| Octave GUI 开发实战 | 于红博 |
| ANSYS 19.0 实例详解 | 李大勇、周宝 |
| ANSYS Workbench 结构有限元分析详解 | 汤晖 |
| AutoCAD 2022 快速入门、进阶与精通 | 邵为龙 |
| SolidWorks 2021 快速入门与深入实战 | 邵为龙 |
| UG NX 1926 快速入门与深入实战 | 邵为龙 |
| Autodesk Inventor 2022 快速入门与深入实战(微课视频版) | 邵为龙 |
| 全栈 UI 自动化测试实战 | 胡胜强、单镜石、李睿 |
| pytest 框架与自动化测试应用 | 房荔枝、梁丽丽 |